Science and Beyond

edited by Steven Rose
and Lisa Appignanesi

BASIL BLACKWELL
in association with
The Institute of Contemporary Arts

First published 1986

Basil Blackwell Ltd
108 Cowley Road, Oxford OX4 1JF, UK

Basil Blackwell Inc.
432 Park Avenue South, Suite 1505,
New York, NY 10016, USA

British Library Cataloguing in Publication Data

Science and beyond.
  1. Science
  I. Rose, Steven   II. Appignanesi, Lisa
  III. Institute of Contemporary Arts
  500   Q162

  ISBN 0–631–14483–8

Library of Congress Cataloging in Publication Data

Main entry under title:

Science and beyond.

  Bibliography: p.
  Includes index.
  1. Science—Philosophy. 2 Science—Social aspects.
  I. Rose, Steven Peter Russell, 1938–
  II. Appignanesi, Lisa.   III. Institute of Contemporary
  Arts (London, England)
  Q175.S416   1986   501   85–18646
  ISBN 0–631–14483–8

Typeset by Cambrian Typesetters, Frimley, Surrey
Printed in Great Britain at The Bath Press, Avon

Contents

List of contributors   vii

1   Introduction   *Steven Rose*   1

Part I   Science at the Limit

2   Biology: A Necessarily Limitless Vista
    James D. Watson   19

3   The Limits to Science   *Steven Rose*   26

Part II   Science, Selection and Sociobiology

4   Structuralism versus Selection – is Darwinism
    enough?   *John Maynard Smith*   39

5   Is Biology an Historical Science?   *Brian Goodwin*   47

6   Sociobiology: the New Storm in a Teacup
    Richard Dawkins   61

7   Sociobiology and Human Politics   *Patrick Bateson*   79

Part III   Science, Brains and Machines

8   Does Artificial Intelligence need Artificial Brains?
    Margaret A. Boden   103

9   Artefactual Intelligence   *Patrick D. Wall and
    Joan N. Safran*   115

10 Minds, Machines and Meaning   *Richard Gregory*   131

Part IV. Towards a New Science

11 Health for All by the year 2000?   *Alwyn Smith*   137

12 Organizing for Science   *Tom Blundell*   157

13 Towards a New Physics   *John Taylor*   161

14 Feminism and Science   *Janet Sayers*   169
15 Nothing Less than Half the Labs   *Hilary Rose*   179
Notes and References   197
Index   203

Acknowledgements

The editors and the ICA would like to thank the International Science Policy Foundation and its Director, Maurice Gold-smith who spurred us to hold the series of discussions on which this book is based on the occasion of the 20th anniversary of the Foundation.

List of Contributors

Patrick Bateson is Professor of Ethology at the University of Cambridge and Director of the Sub-Department of Animal Behaviour. He has edited *Growing Points in Ethology* with R.A. Hinde, *Perspectives in Ethology* with P.H. Klopfer and *Mate Choice*, and co-authored *Defended to Death*.

Tom Blundell is Professor of Crystallography at Birkbeck College, University of London.

Margaret Boden is Professor of Philosophy and Psychology at the University of Sussex and author of *Artificial Intelligence and Natural Man* and *Minds and Mechanisms*.

Richard Dawkins is Lecturer in Zoology at the University of Oxford. He is author of *The Selfish Gene* and *The Extended Phenotype*.

Brian Goodwin is Professor of Biology at the Open University and author of *Temporal Processes in Cells*.

Richard Gregory is Professor of Psychology and Director of the Brain and Perception Laboratory at Bristol University and has written and edited many books including *Mind is Science* and *The Intelligent Eye*.

John Maynard Smith is Professor of Biology at the University of Sussex. His books include *Evolution and the Theory of Genes*, *The Theory of Evolution* and *The Problems of Biology*.

Hilary Rose is Professor of Socal Policy at the University of Bradford and author with Steven Rose of *The Radicalisation of Science* and *Science and Society*.

Joan Safran is Lecturer in Philosophy at the City University in London. Her research interests include the philosophy of language and the philosophy of psychology.

Janet Sayers is Lecturer in Developmental Psychology at the University of Kent at Canterbury and author of *Biological Politics*.

Alwyn Smith is Professor of Epidemiology and Social Ontology at the University of Manchester and President of the Faculty of Community Medicine.

John Taylor is a mathematical physicist and Professor of Mathematics at King's College, London. His books include *Science and the Supernatural*.

Patrick Wall is Director of the Cerebral Functions Research Group at University College, London. He researches on the nervous system, particularly on problems of pain and is author of *The Challenge of Pain*.

James Watson is Director of the Coldstream Harbor Laboratory in New York and won a Nobel prize for his work on the structure of DNA. His books include *The Double Helix* and *The Molecular Biology of the Gene*.

1

Introduction

Steven Rose

Some months ago the Nobel prize-winning immunologist, Sir Peter Medawar, published a slim volume of essays. He entitled it *The Limits of Science*. The essays form a vigorous defence of the boundaries of the empire of science, which he saw as under attack. At a time when critical, anti-science and merely technocratic voices abound, he continues to radiate confidence in scientific rationality, combined with a no-nonsense didactic pragmatism that verges on the heroic. For him, the need is to define science's frontiers (inside culture, but outside politics), to police them against charlatans (like proponents of the genetic basis of differences in intelligence between races) and to define the favoured methodology of 'normal science' as laid down by philosopher Sir Karl Popper. For Medawar, science cannot answer 'ultimate questions' of the meaning of life. Yet within its boundaries, science's search for theories within which to encompass facts is – or can be – limitless. There is always further to go, and science is a good way of getting there.

This confident assertion of the traditional values and boundaries of science is a good starting point from which to consider the issues raised in this book. For the point is that Medawar's optimism and clarity are these days much less widely shared even amongst practising scientists than might have been the case say, 20 or 30 years ago. Then, science was still viewed with almost unalloyed optimism. Scientists, C. P. Snow had told us, were 'men with the future in their bones'.

Science was the bringer of an era of prosperity and leisure, offering a boundless future horizon of expanding human knowledge and welfare. Such was the progressive image, indeed, that the Labour Party under Harold Wilson could campaign and win the 1964 general election under the slogan of 'Building socialism in the white heat of the scientific and technological revolution'.

The optimism about science, and its conjunction with socialism, was not accidental. Wilson himself was echoing the most technocratic aspects of the vision of the Marxist physicist, philosopher and historian of science J. D. Bernal, who saw a vastly expanded science and technology, democratically controlled and centrally planned, at the service of the socialist state, a state which scientific knowledge would play a considerable part in running. To this end, Bernal argued that science itself required studying, as an institution, by way of the disciplines of sociology, economics and philosophy. Ultimately what would emerge would be a 'science of science'. Bernal's vision had dated from the 1930s, but reached its apotheosis with the election of the Wilson government, and in the same year the establishment of a 'Science of Science Foundation' set up to delineate the itinerary along which the Bernalists should march to their goals. The lecture series at the Institute of Contempory Arts from which this book stems marked the 20th anniversary of the establishment of that Foundation, now called the Science Policy Foundation and distinctly more pragmatic in its practice than the grandiose sweep which had been the visionary part of Bernal's agenda.

But there was a further symbolic propriety about holding this series at the ICA. As you enter the Institute's door from the imperial splendour of the Mall, you see on its walls a Picasso mural of a dove of peace. The mural was not originally made for the ICA, however. It had been drawn by Picasso on the wall of J. D. Bernal's house in Albert Street in northwest London, at the time of the 1950 London peace conference. The conjuncture of Picasso and Bernal – artist and scientist – in the politics of peace campaigning was not accidental, for each believed their life's creative work could not be divorced from political struggle. That Bernal's Picasso should end up at the ICA is also fitting, for Bernal, like Picasso – and like Snow and Medawar –

was hostile to any idea of 'two' or even more cultures, of any necessary antithesis between art and science. Both are part of culture, and culture is a seamless web. For Bernal, as a Marxist, any attempt to divide the two would have instantly been attributed to the baneful influence of the bourgeois fragmentation of knowledge, with its insistence on the hierarchy of knowledge from the 'hard' and 'pure' of maths and physics to the 'soft' and 'imprecise' of biology and the social sciences; and the dichotomies of theory and practice, science and technology, mental and manual labour.

To hold a series of discussions on Science and Beyond at the ICA would then seem to bring this particular wheel full circle. But wheels, in a technological universe, are attached to axles and machines; the turning wheel has moved forward as well as round. The prospects which Bernal had viewed, and Medawar still views, so optimistically, are now perceived by most more ambiguously. The public image of science and technology links both with unemployment, not leisure; with weapons of mass destruction, not limitless cheap power; with pollution, not increased health; and with increasing control over and intervention within human life itself, not an extension of human liberty. *Pace* C. P. Snow, it is not so much the future, but radioactivity, which is in the bones. And it's not just the bones of men, but of women and children too. At the same time some of the old certainties about the objectivity of scientific knowledge have been challenged by just those philosophers and sociologists whom Bernal had believed would contribute to the study of the science of science – and even by some scientists too. There are controversies about the range and scope of scientific theories, and about the methods – analytic, reductive, dominating, objectifying – which science is seen as bringing to bear on understanding the world. Far from being timeless, universalist and value-free, scientific methods and knowledge, like the institutions of science themselves, are seen as having embedded within them the values and ideologies of those who are the makers of science – that is, science is seen as overwhelmingly bourgeois, male and white. No longer part of the progressive inevitable march towards socialism, science has become part of the instrumental rationality of capitalism, to use the insight of Habermas. And to an extent undreamed of by the founding fathers to whom I have so far referred, science has been subjected to critical re-evaluation by feminists.

How does this affect the practice of science and the nature of its theories? Not at all, say many of science's spokespeople, echoing Medawar and Snow. Profoundly, say its critics. It is not that the methods of science cannot give us knowledge about the material world. The problem is that the relationship between what scientists *say* about the world and how the world *really* is, is not that simple. Science may hold a mirror to nature, but it is not a plane mirror, rather it is pocked and twisted by the expectations and world view of those who hold it.

If this were not the case, controversies in science would be limited to debates about 'the facts'. All objective scientists, confronted with a given set of facts about the material world, would interpret them in the same way. The task of science would then become one of the mere accumulation of facts, and of developing the technologies with which to acquire more or better facts – to increase the magnification of our microscopes, or the speed at which particles can be whirled in our accelerators, or the resolution of our telescopes – until still more secrets are wrested from nature – to adopt that Baconian image of science which has formed the focus of so much feminist criticism. Most working scientists probably believe that they are in actuality working in this way – collecting facts about the world. But more than a century ago Darwin pointed out quite bluntly that there was no such thing as fact-collecting in a vacuum; facts are always collected *for* or *against* a particular hypothesis. Popper, of course, has made his career as a philosopher of science elaborating Darwin's insight (though without, I believe, ever crediting it, perhaps because of the reservations he has had, at least over a long part of his philosophical career, concerning the scientific status of Darwinian evolutionary theory).

Facts then are collected to be placed in the kaleidoscope of theory, and our perceptions of them are constantly transformed by the shaking of that kaleidoscope. Which brings us to the core of the present book. Rather than rehearse extensively either the external constraints on science, whether financial, political or ideological, or rerun the science/anti-science debates of the 1970s, we have chosen to explore themes of major conceptual controversy in science, where conflicting positions have been taken, where the kaleidoscope is constantly being shaken, the patterns reinterpreted. We have, however, tried to choose themes in which the issues are neither very remote nor internal to a particular scientific discipline itself, but raise matters of profound human concern or social and technological implication. In each case our method has been to invite leading protagonists of particular positions in a contro-versial area to state their case in debate. The draft papers were prepared before the series of debates took place, and were then reviewed by the protagonists in the light of the debate and the discussion which followed each, so each person has had access to the other's text, in the hope that we may thus maximize the light at the expense of the noise of dispute.

Although we have chosen a set of discrete themes, some common threads run through them. An obvious one is that the majority of the participants and themes are drawn from the biological sciences. This is only in part because one of the editors of this book is a biologist! More relevantly it is because the enormous flowering of the biological sciences over the past quarter-century has thrust biology into the centre of the intellectual stage in the way that physics and cosmology were in the period between the 1914–18 and 1939–45 wars. But while issues in physics and cosmology – from the origin of the universe and the ultimate nature of matter to relativity theory and the prospect, aeons ahead, of universal extinction by 'heat death', raised and still raise feelings of profound intellectual wonderment, the challenges posed by biology are to our self-conceptions, our own human place in nature and its relation-ship to the very artefacts we create.

Just because biology – or at least some biologists – claims to address questions about the nature of human nature, the answers biology offers cannot but be charged with ideological significance. Does evolutionary theory imply that certain aspects of human social organization – capitalism, nationalism, the patriarchy, xenophobia, aggression and competition – are 'fixed' in our 'selfish genes'? Some biologists have claimed to answer this question in the affirmative, and political theorists of the Right – from libertarian monetarists to neo-fascists – have seized upon their pronouncements as providing 'scientific' justification for their political philosophies. Are humans 'merely' machines – and if so what becomes of the vaunted free human spirit? Biological questions are thus not merely part of culture, but of politics too.

Such debates within biology have taken on an increased resonance outside the science as well. The collapse of the utopianism of post–1968 dreams, with their romantic belief in the infinite plasticity of a socially constructed human nature, and the decline of any coherent socialist theory grounded in a co-operative concept of the human condition, have opened the door to a new wave of biologically determined human nature theories. These have been the stock-in-trade both of New Right political and of re-emergent fascist groups both in Britain and abroad. Within feminism, debates between biologically essentialist and socially constructionist views of the formation of the sex/gender system have found themselves drawing on biological arguments. The growth of the ecology movement and the increased emphasis on personal health and fitness, too, turn attention towards questions of human biology and its relationship to the natural and social worlds in which we are embedded. The debates within biology thus impinge directly on questions of how we should live and how organize our society.

But this is a two-way street, for one of the other insights of the post-Bernalian and post-Popperian history and sociology of science is the recognition of the nature of those forces which twist and pock the mirror that biologists hold up to nature. If our methods of doing biology, the questions we ask about the non-human biological world, help determine which facts we collect and how we arrange them in our kaleidoscope, then *why* we ask the questions becomes as interesting as the questions themselves. We cannot observe the non-human biological world without bringing to it ideas about the organization of the world derived from our human experience. It is no accident, therefore – and I do not mean this in any conspiratorial sense – that ideas developed in the interpretation of, say human monetarist economics, or the analysis of human sexual relations, find themselves being called upon to explain the behaviour of non-human populations from baboons to Siamese fighting fish, mallard ducks and ants. We cannot avoid this cross-talk; we should at least be aware that it is happening and be conscious of the consequences, which are that what seems like 'pure science' is often pure science filtered through pure ideology.

One further strand runs through the discussions which follow – the issue of the conceptual structures with which scientists, especially biologists – approach the universe they study. The dominant tendency is one which I characterize as *reductionist*. A reductionist methodology argues that the correct way to study complex phenomena is analytical; that is to decompose them into their constituent parts. We decompose societies into assemblies of organisms, organisms into cells, cells into molecules and molecules into atoms. We endeavour to explain 'higher-level' – that is more complex – phenomena in terms of their constituent 'low-level' components. In studying any phenomenon we endeavour to isolate it from the external world, to control all forces impinging upon it, to alter one variable at a time as if by treating each separately we will eventually be able to add the whole set up and reconstitute the dynamic whole once more.

Such a reductionist approach is, methodologically, the way that Western science has developed and proved so successful. For the past 300 years it has provided us with a way of doing experiments which has led to the vast achievements of science as we know it today. But it has proved a method of limited power in dealing with complex phenomena of, say, meteorology, ecology, social interactions amongst animals – or even the human brain and its relationship to the activities of the mind. It is still the case that for all the successes of particle physics, it cannot handle the interaction of more than two particles simultaneously. It is also the case that for all the power of quantum theory, the goal of predicting 'upwards' from the properties of subatomic particles to the properties of molecules has not been achieved even for such simple substances as H_2O.

There are three responses to this impasse: two serious and one trivial. The first is to conclude that reductionism as a *method* is limited in its explanatory power and that as a *philosophy* – that is, as the assumption that it is *the* ultimate way of explaining the world – it is fundamentally flawed. The second is to point to the past successes of reductionism and to argue that it will finally succeed. The trivial response is to deny there is a problem at all, to ignore the debates that the issue of reductionism has raised and continues to raise amongst those who study the philosophy, epistemology and methodology of science as nothing more than a 'storm in a teacup', and to hope that firm language and a dash of healthy Anglo-Saxon empiricism will leap the impasse. For those who hold to one or more of these last two positions – and they are represented amongst the contributors to this book – the critics of reductionism may appear as little better than cranks, mystics, anti-science, or in Watson's memorable phrase, more interested in 'social work' than science. By contrast anti-reductionists see reductionist philosophy as more than just 'bad science', but as reactionary and anti-human as well. Implicitly and sometimes explicitly, this tension is expressed in all the discussions which follow.

The first discussion is intended to set the agenda for all that follows, by confronting the limits to science itself. James Watson, co-discoverer with Crick, Franklin and Wilkins of the double helical structure of DNA in 1953, has been at the centre of molecular biology and its technological exploitation ever since. His teaching books have set the agenda for generations of aspiring molecular biologists and geneticists, and he was the guiding spirit of President Nixon's famous pledge to make the 1970s the decade of beating cancer. Watson, like Crick, is a forthright scientific optimist. For him, science is unproblematic, scientists those who can see most clearly the way ahead, and the task of politicians is to smooth the path of scientific and technological advance. My own position, in debate with Watson, is probably already apparent from the style of this Introduction, and derives from my experience in the so-called radical science movement from the 1960s onwards, from the problems of method I confront day by day in my laboratory practice as a neurobiologist, and above all the experience of more than two decades of close collaboration in all aspects of life, politics and writing with Hilary Rose. I argue, in opposition to Watson, that science is bounded, advertently or inadvertently directed and limited by institutional structures, ideology and the nature of the society in which it is embedded.

The second discussion moves into one of the central debates in contemporary biology, the status of evolutionary theory as a way of explaining the diversity of living forms and their adaptedness to their physical environment. Darwin's *Origins* was published in 1859; what has become known as the neo-Darwinian synthesis was the product of the work of Fisher, Haldane and Sewall Wright in the 1930s and it might seem strange that the status of evolutionary theory is still a matter of debate amongst biologists. But whilst, *pace* the creationists, who sieze enthusiastically on every aspect of the debate as evidence of a 'crisis' in Darwinism, evolution in the sense of the changing of organisms and species with time is a fact, not a theory, the *mechanism* of this change is at present once again the subject of debate. Put at its most succinct, Darwinism pointed out that evolution was an inevitable consequence of the two facts that (a) like breeds like – but with minor variations, and (b) all organisms tend to produce more offspring than survive to breed in their turn. Among the offspring, those more likely to survive must be those which are in some ways more fitted – adapted – to their environment, and their own offspring will tend to be more like them in turn. The consequence is that the characteristics of a population will change over time to be increasingly adapted to its environment. Fisher, Haldane and Sewall Wright offered a mechanism both for the preservation of favoured varieties, and for the preservation of changes in type, by adding knowledge of genetic transmission to produce the neo-Darwinian synthesis.

John Maynard Smith is amongst the most distinguished exponents of what might be called orthodox neo-Darwinism, and has extended its explanatory scope into the fields of the evolution of social behaviour and of sex. Here he offers a terse but magisterial statement of the neo-Darwinian position. But why has neo-Darwinism been under attack? Primarily because whilst it is a very good way of explaining how species get better at doing what they do, it has problems in explaining how new species, doing quite other things, arise. There are other problems too. One is the need to replace the rather sterile metaphor of the environment as the stable 'selector' of the organism, posing challenges which fit organisms 'pass' whilst less fit 'fail', with a dynamic concept of a constantly changing environment and a constant dialectic between organism and environment. A second problem is the need to explain why there have apparently been long periods of evolutionary stasis followed by rather rapid periods of change – the question of what has been called punctuated equilibrium. A third is the continuing issue of what is meant by referring to a particular character as 'adapted', and a fourth that of just what is selected – are genes the units of selection, or are individual organisms – phenotypes – selected. Or is there a selection process acting at the level of the population as a whole?

Brian Goodwin, a developmental biologist and for nearly 20 years a colleague of Maynard Smith's, offers a more radical critique of neo-Darwinism than these, however. For him the debate is over the question – posed in other disciplinary frameworks by structuralism – of whether the forms of organisms are determined purely historically – that is by the seemingly random processes of evolution – or whether there are 'laws of form' which limit the range of available structures on which evolution can act. To take an obvious example, there are good reasons why the science fiction monsters of wasps as big as humans are impossible. The composition of the wasp's body means that, if scaled up many times over its present size, its frame would buckle and its wings be unable to lift it from the ground. For humans it is not merely that we lack the genes to enable us to sprout wings, but that wings capable of enabling humans to fly merely by flapping are structurally not practicable. Goodwin pursues the question of rules of structure in depth, taking as his main example the evolution of the form of the pentadactyl limb. It is not that he in any sense denies Darwinian mechanisms, but that if he is right, Darwinian evolution becomes relatively trivial, a selection not from an infinite *à la carte* menu of possible organisms but a choice amongst the limited range of varieties offered by the structural *table d'hôte*. The important theoretical task is then that of the discovery of the 'laws of form' and the rules by which forms become transmuted. The historical contingencies of evolutionary theory become replaced for Goodwin, by the exact sciences of physics and chemistry.

Sociobiology is an approach to the study of animal behaviour built upon evolutionary premises. Where ethology as a scientific approach to the analysis of social behaviour had been concerned to observe animals in interaction in natural – or nearly natural – communities and to formulate rules governing these interactions, both between adults and their young, sociobiology claims to be offering something new. If all aspects of life were the product of evolution, then behaviour too must have evolved; it must be adaptive to the animal's environment – social as well as natural now – and in so far as it is the product of selection, there must be genes which help 'determine' it. Because natural selection acts so as to preserve genes, rather than phenotypes (the outward form of organisms) each individual animal must act so as to maximize the chances of its own survival – and that of its genes – into the next generation. If this means that in some sense organisms are therefore always in competition with one another, how can acts of social behaviour – so-called altruistic acts – have evolved? Armed with a theory known as kin selection (which points out that as related individuals share some genes in common, it is 'in the genes' interest' for one individual to sacrifice itself if thereby it helps ensure survival of a sufficient number of its kin) E. O. Wilson in the United States and Richard Dawkins and John Maynard Smith in Britain began to develop a theory of animal behaviour based broadly on the claim that all aspects of that behaviour must be looked at from the point of view of the contribution it makes to the propagation of the genes of the organism which manifests the behaviour.

Such sociobiological theorizing is in my view (*pace* Dawkins) intrinisically reductionist, and runs the risk of being bio-logically determinist. The publication both of Wilson's trilogy (*Sociobiology*; *On Human Nature*; *Genes, Mind and Culture*) and Dawkins's own *The Selfish Gene* and *The Extended Phenotype* produced a barrage of comment and criticism, focusing both on the explicit efforts made by Wilson in particular, but also Dawkins in a much-quoted passage on the dangers of the welfare state blunting the edge of the Darwinian imperative, to derive precepts for human action from the supposed secure base offered by sociobiology, but also for their implications for biological theory making. My own *Not In Our Genes*, written jointly with Richard Lewontin and Leo Kamin, is a reflection on some of these issues and a critique of sociobiology. Dawkins, invited to debate the theme of 'where does our behaviour come from?' with Patrick Bateson, chooses to do so in part in terms of a critical response to *Not In Our Genes*, and his chapter must be understood in that context. He is concerned to advance the claims of a sociobiology distanced somewhat from E. O. Wilson's, to rebut the charge of biological determinism, and, simultaneously both to defend reductionism in principle and deny that his approach is philosophically reductionist in practice.

Patrick Bateson, whose work derives from the more classical ethological tradition (and whose research has been directed primarily towards uncovering ways in which early experience in the young animal and its first interactions with parents, siblings and other conspecifics affects its subsequent behaviour when adult) responds to Richard Dawkins at a number of levels. He is not unsympathetic to the project of sociobiology, at least in Dawkins' as opposed to Wilson's hands. It is not that Dawkins is a reductionist or a determinist, says Bateson, it is just that his powerful writing style and arresting use of metaphors makes it *seem* on occasion as if he is. But there are points of difference which Bateson seeks to bring out. Where for Dawkins it is in the last analysis only meaningful to talk of selection acting at the level of the gene (the gene is 'the unit of selection'), for Bateson selection must be seen to be acting at a multitude of levels; at the gene, certainly, but also at the level of the entire genotype, or ensemble of genes that each organism possesses; at the level of the organism as a whole (its phenotype) and indeed of populations of those organisms. Further, where Dawkins stresses competition as the motor of evolution, Bateson is concerned to point out that competitive mechanisms can result in the evolution of co-operative behaviours. Here, he follows – consciously – a tradition deriving from the nineteenth-century anarchist theorist Kropotkin. Bateson concludes by drawing the appropriate lessons for human politics.

The fourth discussion turns to another powerfully emotive area of current debate in biology, psychology and philosophy. Do the rapid advances in computer technology – leading

towards the ambitious Japanese Fifth Generation project –
bring us any nearer the possibility of artifical intelligence (AI),
of making computers which work like brains, or robots which
can behave like people? It is not merely that this has been a
recurrent theme in science fiction over many decades now, but
vast sums are being invested, in Japan, the United States, and
now belatedly in Britain, into information technology and
artificial intelligence projects. The 1984 Reith lectures, by John
Searle, gave one philosopher's views as to why AI was a
contradiction in terms. By contrast, Margaret Boden, as a
philosopher who has worked extensively in the history and
philosophy of psychology, thinks otherwise. Being brainy, she
argues, is not contingent on being constructed of flesh and
blood, of neurons and synapses, but of being able to manipulate
logical constructs, act cognitively and handle and develop
concepts. To believe otherwise is to accept a metaphysical
view of human nature which is antithetical to science *and*
philosophy.

The opposing case is put by Patrick Wall and Joan Safran. As
a neurophysiologist whose work has concentrated primarily
on the study of the physiology of pain, Wall's response is
grounded in the nature of real brains as they can be studied
observationally and experimentally, but is enriched by the
input of the philosopher Safran. Computers, they say, are
analogies to brains. But what sort of analogies? They are not,
contrary to Margaret Boden, organized conceptually or opera-
tionally along similar lines to brains; rather they are at best
evocative analogies. What is more, for all the grandiose claims
of the computer buffs, they are in practice not aiming at
intelligence in the human sense, nor at producing models of
brains in the sense which will help us understand how real
brains work. This is because real brains are the products of
evolutionary and developmental history, a history of contingencies which the computer cannot match. Asked to produce
an 'intelligent' machine, and faced with a real human brain, a
computer engineer is likely to respond 'I wouldn't start from
here'. In a brief summary statement concluding this discussion,
psychologist Richard Gregory – whose own work is a constant
attempt to bridge the gap between what machines can do and
what human perception is and is not capable of achieving –

offers a mediation between these contrasting positions in terms of the meaning of meaning.

In moving from evolutionary theory to the artificial intelligence debate we are gradually turning towards the social outputs of modern science in terms of technological developments. The final five chapters continue and extend that transition, focusing on the theme 'the future of science'. The wonders of modern biology, hinted at by Watson, are beginning to generate new possibilities for medicine and health care. The present debates over genetic engineering, gene therapy and *in vitro* fertilization (test-tube babies) are but one manifestation of a range of technological potentials now opened up. Does our future health depend on the extension of these technologies to the maximum, or are the major determinants of health and disease to be found not inside individual biology but in the social environment?

Alwyn Smith, who is one of the central and charismatic figures in the development of community medicine in Britain, points to the relatively minor contribution of medicine to the achievement of health. Mortality has dropped over the last century and life expectancy risen, but these changes are not in any obvious way connected with improvements in basic biological knowledge or medical technique. Today's major killers, coronary heart disease and cancer, present a new set of problems which again, despite Watson's view of the importance of molecular biology for the solving of the cancer question, seem to require epidemiology and changes in personal lifestyle (like smoking) or social structure rather than 'high technology' medicine.

Tom Blundell, who not only occupies Bernal's old chair of crystallography at Birkbeck College but shares many of his concerns, is more positive about the role of science. We do not invest enough, he argues; we do not control and plan our science properly and we need to open up and democratize decision-making structures. Most optimistic is John Taylor, a mathematician and physicist with great capacities for enthusiasm – an enthusiasm which took him a few years ago into the uncharted waters of parapsychology and spoon-bending but are now devoted to what are for some of us nearly as esoteric areas of the new physics. Sharing the optimism of

Watson and Dawkins, perhaps even that view of the 'golden age' of Gunther Stent to which Watson refers – he looks forward to the successful grand unified theory of physics, with which the world will be explained and made whole again.

Finally we turn to the voices most conspicuously absent from the discussion so far – those of feminism. Over the past decade feminism has begun to develop a powerful critique of male science. It is not merely that women are systematically excluded from the ranks of those who male scientists define as 'the scientific community', but biological determinist theories seek to give this exclusion, the 'inevitability of the patriarchy', an air of necessity. At the same time male science, as it is practised, seems to be a science of domination, of violence, of death. Both Janet Sayers, a psychologist, and Hilary Rose, a sociologist, have made major contributions to the feminist critique of science. Janet Sayers's book *Biological Politics* looks particularly at the ways in which nineteenth-century science portrayed women, and its sexist preoccupations. Here, she looks at the systematic sexism of science in its practice and methodology, and shows that scientific theory-making is not neutral, through observing the social history of the management of childbirth. Finally she seeks to root these patriarchal practices in the social division of labour, between production, a male concern, and reproduction, a female concern. Is it merely utopian to ask for a new, a feminist science?

This is Hilary Rose's central preoccupation, in the final chapter of the book. She points first to the fact that women are present in science – but invisibly, in the subordinate roles of technicians and cleaners and secretaries. To bring more women visibly into science and technology will profoundly change the nature of that science and technology. The new social movements, of feminism, ecology and peace, actively seek a new relationship with nature, a relationship which transcends the dichotomy of objectivity and subjectivity and looks to holistic rather than reductionist, harmonious rather than dominatory, ways of living with and understanding the world. Such types of knowledge, which unite not merely hand and brain, but, to quote one of her earlier papers, 'hand, heart and brain', are the contribution of a feminist epistemology which derives not from some type of biological essentialism, but from a division of labour which assigns to women the caring tasks of our society. Whereas the old Marxist Left argued that the choice for the world was between socialism and barbarism, for Hilary Rose the choice – both for science and humankind – is between transcending male values and moving towards a feminist epistemology – or destruction. As a modest step in this direction, she proposes the slogan 'nothing less than half the labs'. For her, and for feminism, this is the starting point for a new science.

Part I Science at the Limit

2

Biology: A Necessarily Limitless Vista

James D. Watson

The trouble with a discussion of 'the limits' to any science is that it lays one open to the temptation to engage in verbose generalizations and to set out vast theoretical schemes. Even worse, it slightly opens the door to philosophy. I do not like to suffer at all from what I call the German disease, an interest in philosophy. In the early nineteenth century the Germans invented *Naturphilosophie* and still today there exist other-wise quite intelligent scientists who worry about the philo-sophy of the cell and so forth. Some academics may find this an interesting way of looking at things, but it has never come up with interesting predictions, and I have an abhorrence of abstractions that lead nowhere.

My theme is obviously to be whether there are or should be limits to biology rather than to science in general; but I should begin by saying what I mean by science. It is simply a method of trying to explain the animate and inanimate objects around us – or at least their reproducible parts; how the world works, and what it is made of. That is, what are sometimes called the laws of nature. What is more, I am not proposing to speculate about what might happen over the next 100 years, but will confine myself to the next 10–20 years and what is possible in that time. I look forward to these years with much excitement since we have reason to believe that the pace of discovery will be maintained. There is no evidence at all of any slowing down in biology due to an inherent limitation in the meaningful facts that can be learned in the conceivable future. An opposite view nonetheless was put forward semi-seriously after we found the structure of DNA and solved the genetic code. Since those triumphs marked the end to a very long intellectual quest, Gunther Stent, in his book *The Coming of the Golden Age*, argued that with such knowledge, and that which would soon follow, the world might commence an age of tranquillity and affluence since the historic phase of human striving for knowledge would be at an end. For myself, I have never accepted that the DNA structure and code were such ultimate solutions that soon there might be nothing important left to do. Now, especially in this recombinant DNA era, we see DNA as a beginning, not an end.

Neither am I really interested in whether it can be argued that in the past 35 years we have done better science than we will in the next 35 years (physicists over 70 often reminisce that physics has never had it as good as when quantum mechanics was young, say in the early 1930s). I do not think it is worth worrying about – I have told my young son (14) that if he wants to have an exciting career in science he should try to find out how the brain develops and works. He need have no fear that it is going to be even slightly solved before he could learn enough (say in 10 years) to be able to think deeply on his own. Even when you think someone has taken the really big step forward and climbed a very tall intellectual mountain, within several months you usually see some further peak that you want to climb to feel satisfied. The double helix made me content for at most 6 months before I began to feel itchy and bored with our lack of immediate further progress.

A question I find more interesting than the abstract discussion of the limits to scientific knowledge is how far you can limit human curiosity? Are there ways of prescribing what humans think about? There are two considerations here. One argument is that what we are doing is dangerous and therefore we should not be doing it. The second is that, even if we should not find out certain things, but nonetheless we do, how long can these dangers be contained? These two considerations are of different types: one is moral, the other is a question of physical and legal constraints. But whose task should it be to set these constraints, if indeed there should be any?

Until 12 years ago these were not very real issues for most biologists. Then suddenly recombinant DNA procedures gave us the power for genetically engineering life forms in a semi-predictable as opposed to random manner. Almost from the start of this era we saw its immense possibilities for creating genetically altered bacteria, if not higher plants and animals. So, many leading molecular biologists asked whether their science might slip out of control and produce new forms of life, if not knowledge, that we would not be able to cope with. If so, we had to create procedures where the people working in molecular genetics would help set up a rational system of regulation. This mood led to a 1974 letter to *Nature* and to *Science* from the molecular biologist Paul Berg and several other colleagues, myself included, calling for a conference to discuss the question of whether and how we should limit what types of experiments could be done.

However, when we met at Asilomar on the California coast in 1975 there turned out to be no way we could either quantify the dangers or logically plan the future. Nonetheless, the mid-seventies was a period when it seemed especially the right thing to put the apparent interest of society as a whole in front of our own personal goals. So with the approval of virtually all concerned scientists, containment facilities and governmental committees to vet experiments grew out of the Asilomar conferences. At first these decisions were widely applauded even among scientists whose experiments had become restricted. Then it became clear that regulations without facts had to be arbitrary and, when put into uninformed hands, could badly block our science. One key field so prevented were studies aimed at the isolation of genes that cause cancer. So we fought back and, within a decade, many of the worst regulations are now gone. Such a withering away was our rational response to the fact that we have no way of telling in advance whether the modified organisms can harm us.

Are we now better off as a society because we raised the issue and tried to reassure the doubts so generated? The various containment procedures we adopted, however, never satisfied the most vocally persuasive opponents of DNA work that the newly created, possibly pathogenic, organisms would not escape from our labs and cause epidemics or run out of control in some other way. You could spend 10 million or 100 million dollars and still not know if you had done enough. The most commonly accepted viewpoint now is that the Asilomar experience was necessary for us to go through because it sent the message that the scientific world was prepared to consider the future good of mankind more important than the specific scientific goals of the immediate moment. Now I am not so sure we did the right thing. We have only so much psychic energy for worry, and thus society as a whole might be better off if all this emotional energy had been devoted, for example, to working toward a safer approach to nuclear weapons.

This consideration brings us to a related issue. Should we go ahead with research which, unlike virtually all recombinant DNA research, actually has a true smell of danger? Consider tumour virology. Many labs study viruses that cause cancer in animals and which also might be the cause of some human cancer. Given the long incubation time of many cancers to develop, we have no way of knowing when we start working with a new such virus whether we are putting ourselves or our technicians at significant risk. Nevertheless, it seems more moral to go on with, as opposed to avoiding, such research since it is likely to yield important facts that will let us cure or prevent future cases of cancer. We have thus tried to create labs which minimize the risk of potential infection without spending so much money on safety factors that none is left to learn the essence of cancer. Whether we have in fact made the right decisions will only be known many years, if not decades, into the future.

In the end you have to be optimistic that human intelligence can rise to the unknown, even when the new challenges have no immediate easy solution (e.g., AIDS). Among scientists, only optimists succeed since a pessimist will not go into unknown territory. Important questions can only be solved by going where no one has been before. I see no other way to advance. This is the way Western civilization has developed, and if we are to survive further as a civilized force, we must continue to face the unknown. This means we will have many scares ahead of us, but not necessarily fewer than if we tried to block further new facts to think about. Something like AIDS is bound to emerge periodically and force us to acquire new

remarks seemed like Hitlerism, denying the essence of humanity that must be preserved at all cost. Like most of Francis's opinions, however, I believe it was very far sighted. We would have a far better society if we accepted that we are the imperfect products of evolution, not of a deity whose judgements we cannot question. I cannot see any good at all coming from letting an infant live a necessarily totally sick and wretched life, no matter how short the period. But since there still is much disagreement as to what forms of living are meaningful, I see little chance that ours or any other society will soon adopt Crick's proposal.

So, in too many situations we still will have no choice as to whether we will become the parents of children horrendously disadvantaged to face their lives ahead and often virtually condemned to a doomed existence (e.g., children who are born with a bad muscular dystrophy). Now, fortunately, an increasingly greater number of parents, thanks in large part to the new ante-natal diagnostic techniques that are being developed using recombinant-DNA-based procedures, will have a choice to abort such genetically damaged fetuses, hoping that their next conception will lead to a fetus that will have a much better chance of developing into a totally normal child.

In the end, our real problem as scientists is not deciding what forms of biology we stop or limit, but in coming to grips with what we find out about the world of living creatures and acting rationally and compassionately on the basis of what we discover. If we behave in such a manner, then the lives of our children, and even more so of their children, will have a good chance of being substantially more rewarding and enlightened than the lives of those of us who toil today.

3

The Limits to Science

Steven Rose

For the great ideological 'spokesmen' of science, from Francis Bacon to James Watson, science has always been without limits; about 'the effecting of all things possible'. Human curiosity, after all, is boundless. There seems to be an infinity of questions one can ask about nature. At the end of his long scientific career Isaac Newton felt, he said, as if he had merely stood at the edge of a vast sea, playing with the pebbles on the beach. What is more, because science is not merely about passive knowledge concerning nature but about the development of ways of changing nature, of transforming the world through technology, these same apologists offer us a breathtaking vision of the prospect of a world, a nature – including human nature – made over in humanity's image to serve human needs. It is only when one looks a little more closely at these visions that one sees that a science which claims to speak for the universality of the human condition, and to seek disinterestedly to make over the world for human need, is in fact speaking for a very precise group. Its universalism turns out to be a projection of the needs, curiosity and ways of appreciating the world not of some classless, raceless, genderless humanity, but of a particular class, race and gender who have been the makers of science and the framers of its questions indeed from Francis Bacon onwards.

The ideology is powerful, and in the second half of this century has been of endless fascination to politicians as well as scientists. Towards the end of the Second World War, in the

USA, Vannevar Bush, whose life had been spent with '*Pieces of the action*' of science, offered Presidents Roosevelt and Truman '*Science, the Endless Frontier*' as a vision of how the greatness and power of the USA, could be indefinitely extended. In Britain the visionary Marxist tradition of J. D. Bernal inspired Harold Wilson in 1964 to speak of the 'building of socialism in the white heat of the scientific and technological revolution' and Soviet scientists and politicians to speak of the 'scientific and technical revolution' which has, rather than politics and class struggle, become the motor of the growth of Soviet society.

Against Watson's claims for the limitless nature of human curiosity and the techno-enthusiasms of the politicians, the anti-science movement of recent decades has cried a series of halts. Halts to the 'tampering with nature' of the nuclear industry and militarism; halts to the possibility of knowledge by the endless dissection of animals into molecules, and molecules into elementary particles; halts to the restless experimentation implied by the very scientific method itself as a way of knowing the universe, as opposed to the contemplative knowledge offered by alternative philosophical systems.

I am not an anti-scientist in this, or indeed in any sense that I would accept. I want to argue, however, that we cannot understand science or speak of its limits or boundlessness in the abstract. To speak of 'science for science's sake' – as if, to paraphrase Samuel Butler on art, science had a 'sake' – is to mystify what science is and what scientists do. This mystification, still often on the lips of the ideologues of science, serves to justify specific interests and privileges. Instead, we have to consider *this* science in *this* society. I shall argue that it is indeed limited, and that its limits are provided by a combination of two major, though only partially separable, factors. The first is material, the second ideological. I will consider each in turn.

The material factor is of course that of resource. Science costs money, and in the advanced industrial countries of Europe, East and West and the USA, consumes anything from 2 to 3 per cent of GNP. From 1945 to the late 1960s, science was expanding at an enormous rate, an exponential growth with a doubling period of 10–15 years or so. A historian of science,

Derek de Solla Price, pointed out that the doubling rate had been constant from about the seventeenth-century on. However, like population growth, scientific growth could not continue unchecked. It became fashionable in the 1960s to calculate that by the twenty-first-century every man, woman, child and dog in the world would be a scientist and the mass of published research papers would exceed that of the earth.

Something had to stop, and indeed it did; from the late 1960s on, in most countries, the growth of science as a proportion of GNP slowed, halted or was even, in Britain, reversed. Sheer resource limitations were limiting the growth of science. You can see this in the development of the physicists' accelerators. First, each country had its own. Then there was the West European CERN project at Geneva, and matching machines in the USA and USSR. Now, even if Britain were to stay in CERN, which is at present doubtful, the costs of the new generation of machines make the 'world accelerator' the logical next step. And beyond this? Just how much resource is going to be devoted to whirling particles around at speeds closer and closer to that of light? Boundless human curiosity is going to be bounded.

Of course the particle acclerator episode is revealing in another way. Ask high-energy physicists why anyone should spend hundreds or thousands of millions of pounds on them, and you are likely to get the answer that it is high culture, and surely society can afford it, like subsidizing Covent Garden. But they will be fooling themselves – or you; because it is not the story they have been telling the politicians, who have gone on shelling out vast sums of money for physics since Hiroshima and Nagasaki in the not unreasonable belief that they would get bigger and better bombs, or new sources of power, out of the investment.

This brings us to the more important point about the material limits on science; for funding is not merely limited: it is *directed*. Of the 2–3 per cent of GNP Britain has spent on science since the 1950s, getting on for 50 per cent, year in, year out, has gone on military research. The figure is now about 53 per cent – the highest for many years, and much more, incidentally, than is spent by any other Western country except the USA – compare France's 35 per cent, Germany's 12 per cent and Japan's less than 5 per cent. If you want to know why so much scientific endeavour is directed to military ends, you must ask political questions about how the decisions are made. There can be no doubt, however, that this concentration on directing research towards military needs, and towards the industrial priorities of production and profit, as Hilary Rose and I have described it, profoundly shapes the direction in which science goes.

Apologists for the purity of science (though it is the purest of high-energy physics that gave us the bomb) may argue that this is all technology – real science is unaffected by such directive processes. They are on shaky ground, making this science/technology distinction, of course. The distinguished American organic chemist, Louis Fieser, invented that nastiest of conventional weapons, napalm, experimenting on it in the playing fields of Harvard during the 1939–45 war. He wrote about his discovery afterwards in a fascinating book called simply *The Scientific Method*. The argument that pure science is divorced from direction cannot be sustained for a moment.

Take the triumphant progress of molecular biology these past decades. There have always been two broadly contrasting traditions in biology: a reductionist, or analytic and atomizing one; and a holistic or more synthetic one. This latter tradition – and I will have more to say about both in a moment – was strongly represented in the 1930s by such developmental and theoretical biologists as Needham, Woodger and Waddington. There was a proposal to set up a major institute of theoretical biology in Cambridge which would have brought the field together; but the funding agency was to be Rockefeller, and Rockefeller, under the guidance of Warren Weaver, decided that the future was to be chemical. They backed biochemistry and molecular biology instead. The double helix and all that followed from it from 1953 on was a direct result of that funding decision. Many people would argue it was a correct one, and I might well agree. The fact is that it changed the direction of biology by a deliberate act of policy. Rockefeller's decision is thus comparable to those being made routinely by government and charitable funding agencies as they decide which are high-priority areas to back, and which should not be supported. MRC, SERC and the rest have their priorities. They are, as it happens, still mainly molecular, even though most of the problems which MRC ostensibly exists to help solve are clearly not going to be resolved by more molecular biology. One of the things that is clear from the combined efforts of Richard Nixon and Jim Watson in the 1970s to 'cure' cancer by the end of the decade is that the most exquisite molecular biology has brought us no nearer controlling a disease many of whose precipitating causes are located in the chemical environment of our industrial society. The vast funds Nixon allocated *have* given us more and more molecular biology, though.

Let me move from the material to the ideological limits to science. The point I want to make here is not just that we get the science we pay for, but that at a deeper level, what science we do, what questions scientists consider important and worth asking at any time – indeed, the very way they frame the questions – are profoundly shaped by the historical and social context in which we frame our hypotheses and realize our experiments. Let me spell this out at three levels:

First, we can only ask questions we can begin to frame; the question of the role of chromosomes in cell replication and genetic transmission was unaskable until there were microscopes powerful enough to see the chromosomes, as well as a genetic theory to be tested. The technology and the theory came together at the beginning of the present century.

Second, not all scientific facts are of equal value. There is an infinity – in the strict sense of the term – of questions one can ask about the material world. Which ones are relevant at all, is strictly historically contingent. To give an example, in 1956 Sanger published the complete amino acid sequence of a protein, the first time anyone had done it. It took him about 10 years. The protein was insulin, and he got the Nobel prize for sequencing it. That it was insulin, rather than any of the other 100,000-odd human proteins, or the thousands of millions of other naturally occurring proteins, was fortuitous. It happened to be a relatively small molecule and available pure and in bulk. Within a few years several other proteins were sequenced, each time to a great, but diminishing scientific fanfare. Today anyone can do it within a few weeks with an automated machine. But is anyone going to *want* to determine the structure of *all* naturally occurring proteins – or even all human ones? There is a law of diminishing returns, to all except stamp collectors, and sometimes PhD students. So a new fact – the sequence of another protein – is nothing like as interesting as the first protein facts were. There is a limit to how many such facts are wanted, and most protein sequencing projects are scarcely worth a research grant these days.

Third, and at a much deeper level than either of the two previous points, there is, it seems to me, a fundamental limit to the capacity of science, framed within the dominant paradigm in which most of us work to give meaningful – let alone satisfying – answers to the great questions of human concern today. The issue of reductionism and its alternatives runs like a thread through many of the discussions in this book, so I will do no more than sketch the issues here. I have written more fully on the subject of reductionism elsewhere (e.g. *Not in our Genes*, and *More than the Parts*). The point is that the mode of thinking which has characterized the period of the rise of science from the seventeenth-century is a reductionist one. That is, it believes not merely that to understand the world requires dissassembling it into its component parts, but that these parts are in some way more fundamental than the wholes they compose. To understand societies you study individuals, to understand individuals you study their organs; for the organs their cells; for the cells their molecules; for the molecules their atoms . . . right down to the most 'fundamental' physical particles. Reductionism is committed to the claim that this is *the* scientific method, that ultimately the knowledge of the laws of motion of particles will enable us to understand the rise of capitalism, the nature of love or even the winner of the next Derby. It also claims that the parts are ontologically prior to the wholes they compose.

The fallacies of such reductionism should be apparent. We cannot understand the music a tape recorder generates simply by analysing the chemical and magnetic properties of the tape or the nature of the recording and playing heads – though these are *part* of any such explanation. Yet reductionism runs deep. For Richard Dawkins the well-springs of human motivation are to be interpreted by analysis of human DNA; for Jim Watson 'What else is there but atoms?' Well, the answer is –

the organizing relations *between* the atoms, which are strictly not deducible from the properties of the atoms themselves. After all, quantum physics cannot even deal with the interactions of more than two particles simultaneously, or predict the properties of a molecule as simple as water from the properties of its constituents. Think of a Martian coming to earth and being confronted with the parts of an internal combustion engine. What are they for? The parts do not make sense by themselves; not even when they are reassembled into a car, unless you know as well that the car is part of a transportation system.

Yet why do scientists of experimental ingenuity and reputation consistently claim that you can understand the transportation system from the parts of the car engine? The roots of that belief go back, I think, to the Newtonian and Cartesian project for science as it has developed from the seventeenth-century, and, in ways I have not space to elaborate here, are profoundly interconnected with the process by which north-western Europe gave birth in the mid-seventeenth century not to a single child, science, but to the twins of science and capitalism, whose growth has subsequently been inextricably intertwined. Reductionism was a scientific philosophy customized for capital's needs, and has remained so since. The trouble is that just as capitalism was once a progressive force but has now become profoundly oppressive of human liberation, so too with reductionism. Beginning as a way of acquiring new and real knowledge about the world – from the structure of molecules to the motions of the planets – it has become an obstacle to scientific progress.

So long as science – in the questions it asks, and the answers it accepts – is couched in reductionist and determinist terms, understanding of complex phenomena is frustrated. A reductionist science, I believe, cannot advance knowledge of brain functions, or solve the riddle of the relationship between levels of description of phenomena such as the 'mind–brain problem', which Western science is almost incapable of conceiving except in Cartesian dualist or mechanical materialist terms. Reductionism cannot cope with the open, richly interconnected systems of ecology, or with integrating its scientific understanding of the present frozen moment in time with the dynamic recognition that the present is part of a historical flux, be it of development of the individual or of evolution of the species.

Failing to approach the complexity of such systems, reductionism resorts to more or less vulgar simplifications which, in the prevailing social climate, become refracted into defences of the *status quo* in the form of biological determinism, which claims that the present social order, with all its inequalities in status, wealth and power, between individuals, classes, genders and races, is 'given' inevitably by our genes. This limit to the scientific vision is compounded by the closed recruitment process into science as an institution which effectively ensures its preservation as the privilege of the Western white male. However, referring to *that* limit to science extends my agenda here further than space allows. I want to conclude by referring to the one limit to science I have not yet mentioned, and that is the *ethical* one.

Ethical issues in science have been repeatedly discussed in recent years. They take several forms. On the one hand, some claims have been made that certain types of knowledge are too dangerous for humanity in its present state, and therefore some types of experiment should not be made. For instance, nuclear power, or gene cloning, are considered to present hazards which make it inappropriate to pursue them experimentally. Or research on the so-called 'genetic basis of intelligence' might reveal biological 'facts' which would be unpalatable. On the other hand, it has been argued that the conduct of certain types of experiment – for instance, those which cause pain to animals, or for that matter to humans – contravene absolute moral principles and should not be performed. All of these considerations may be regarded as limiting science.

From what I have already said it should be apparent that I have a complicated response to that rather abstract approach to ethics. For me, the resource and ideological questions are paramount, and most ethical questions eventually break down to ones about priority and ideology. For instance, there has been a lot of attention given to the ethics of *in vitro* fertilization – should we or shouldn't we? To me, the question seems wrongly posed; instead, one should ask the prior question, which the *in vitro* fertilization techniques are presumably designed to help answer: how can we increase the number of wanted, healthy babies? If I ask that question, I also begin to ask what prevents wanted, healthy babies surviving; and I note that in Britain the perinatal mortality rate – that is, the number of babies dying at or just after birth – is much higher in certain geographical areas – for instance, Liverpool – than others – such as, for instance, Hampstead. I note that there is a severalfold greater chance of a baby not surviving if it is born to a mother in poverty, or in the manual working class, than if it is born to a wealthy or upper-middle class mother. So if we want to save babies, I conclude, we can do so best by applying known social, economic and health care improvements to deprived geographical areas and classes in Britain. *In vitro* fertilization is a method which is of relevance to a small number of relatively privileged mothers. The language of priorities says that we should not get excited about that new set of techniques until we have addressed the question of how we save babies we *know* statistically will die from lack of application of quite simple preventative and health care measures.

That is an ethical question, certainly, but it is also one about politics and economics. Personally, I would not do research funded by, or with obvious applications to, the military. I will try to persuade as many of my fellow scientists as possible to take a similar ethical and political decision. But in the last analysis in a militarist society *anything* one does can be, and potentially will be, co-optable for military purposes. If we do not want war-oriented research, individual ethical decisions are not enough. We need the *political* decision not to finance war research.

Similarly, I accept the case made by the animal liberationists that it is undesirable to use procedures likely to cause pain or distress to animals – though in the last analysis I owe my prior loyalty to my own species, and to argue otherwise seems perverse. I care more about saving people than saving whales. But a vast proportion of the animal experimentation done in Britain is either for relatively trivial commercial purposes – for instance, developing new drugs when it is at least arguable that there are enough or even too many drugs available already. What is needed is not new magic drugs but a health-producing society. Indeed many drugs which are developed and tested are not new in a real sense, but part of the endless process of molecular roulette played by the drug companies in their efforts to circumvent patent laws or maximize profits. It is also true that a fair number of the animal experiments done in 'basic science' labs are, on close analysis, carried out in the pursuit of trivial or 'me-too'-type research aims. Remember that the average scientific paper is probably read by only one or two other people apart from the editor of the journal in which it appeared and the referees. So part of my answer to the question of ethics and animal experiments is to rephrase the question in terms of whether the research is worth doing anyhow, animals or no.

So too with the question of 'thing we are not meant to know'. These are often just things it is not worth trying to know – like the sequence of every possible naturally occurring protein that I referred to earlier. Sometimes, however, they are things which *cannot* be known because the questions are simply wrongly or meaninglessly phrased. As someones who has been involved in what has become known as the 'race–IQ' debates, I have often been asked whether I am opposed to work on 'the genetics of average race differences in IQ' on ethical grounds. My response is that I am opposed to it on the same grounds that I am opposed to research on whether the backside of the moon is made of gorgonzola or of stilton. That is, it is a silly question, incapable of scientific answer and actually, *sensu strictu*, meaningless. The question makes grammatical, but not scientific, sense, because 'IQ' is not a phenotype susceptible to genetic measurements and heritability estimates cannot be applied to average differences in phenotypes between groups.

All this is not to duck the questions of ethics. There are issues of real choice and dilemma in medicine, in the use of animals, and indeed in some aspects of biotechnology, which cannot simply be reduced to issues of economics and ideology. They are few, but important, and they set limits to our science. How should they be resolved? In the last analysis, it seems to me, not by scientists playing god-in-white-coat and refusing to allow anyone else in on the decision. And not by committees of professional ethicists and philosophers. The only way of dealing with such issues is by democratic participation in the decision-making about what science is done. My own aim would be for a way of controlling and directing research which opened all laboratories up to community involvement in their direction, and planned work by a combination of the tripartite structure of decision-making by scientists and technicians in the lab itself, by the community in which the lab was embedded and by discussion of overall priorities and resources at a national level. I believe that if we did organize our science in this way, not merely would new priorities set different limits to our work, but we might also begin to see the makings of a new, less reductionist and more holistic, human-centred science.

Part II Science, Selection and Sociobiology

4

Structuralism versus Selection – is Darwinism enough?

John Maynard Smith

As I shall explain later in this chapter, I do not think that Darwinian natural selection is the only thing we need to understand if we are to understand evolution. However, I do think that Darwin's theory is correct, and that it is the only adequate explanation for what is for me the most characteristic feature of living organisms. This feature is the way in which their structure and behaviour adapts them to survive and reproduce in a specific environment. I never cease to be astonished by this phenomenon. Let me give an example. Recently, for the first time, I had the good fortune to be in Crete when the spring flowers were at their best. Inevitably, I was fascinated by the orchids. I knew already that some orchids resemble a bee visiting a flower, and in this way attract male bees which will transfer their pollen. I knew also that the flowers produce a substance which mimics the sexual attractant produced by female bees. What I observed for the first time is that the flowers of one species have an additional feature which makes them look like an insect. Seen from the side, the lip of the flower has the characteristic three-lobed appearance of a bee or fly – head, thorax, abdomen – and on the side of the head lobe there is a red ellipse, just where the red insect eye should be.

It may be that this simple observation had a special impact on me because I have spent many thousands of hours looking down a microscope at the geneticist's favourite fly, *Drosophila*, and am therefore particularly familiar with the gestalt of a red elliptical eye and a three-lobed silhouette. My point here, however, is that natural selection is the only plausible explanation for this kind of fit between an organism and its way of life. The Lamarckian concept of the inheritance of acquired characters is, so far as I know, the only alternative scientific hypothesis capable of explaining adaptation. Even if characteristics acquired during an individual's lifetime did affect the nature of that individual's offspring – and most geneticists hold that they do not – I do not see how an elliptical eye spot could arise as a response to the environment during an individual lifetime. The natural selection of variants which are non-adaptive in their origin is, I think, the only tenable explanation.

Natural selection, however, can only act on those variants that arise in the first place. Although orchids look like bees and smell like bees, they do not buzz like bees, presumably because a mutant that buzzes has never arisen. It is sometimes said – usually by critics of Darwinism – that mutation is random. Now 'random' is a notoriously difficult word to define. I think that most scientists, when they speak of an event being random, mean that it would not be efficient to enquire into its causes, either because they think that the cause is in principle unknowable, as in quantum theory, or because it would be too much trouble to discover. In this sense, mutation is certainly not random. A lot is known about the causes of mutation – that is, of changes in DNA. However, most geneticists do hold two things to be true of mutation. First, there is no restriction on the kinds of changes in sequence of DNA molecules that can arise by mutation, any more than there is a restriction on the sequence of letters that can be produced by a typewriter. Second, if a mutation is caused by a particular agent – for example X-rays – it is not in general true that the effect of the mutation will be to make the organism that carries it more resistant to the causative agent: in brief, mutations are not adaptive. However, this unrestricted nature of mutation at the DNA level is quite consistent with the view that only certain kinds of structural change are possible to particular organisms, just as only particular sequences of letters make sense in a given language.

I first met the idea that each kind of organism produces a restricted range of variation from my teacher, Helen Spurway, who, in 1949, published a paper entitled 'Remarks on Vavilov's law of homologous variation'. In it she gave many examples of such restrictive patterns of variation. For example, for reasons that I do not understand, mammals go in for horns; they occur in seven of 25 orders, including the rodents. In contrast, birds never have horns, but often have naked patches of skin on their heads and necks – combs, wattles and so on. Spurway then remarks 'These possibilities of mutation or variation determine the evolutionary possibilities of the group.' She does not claim originality for this idea: she quotes T. H. Huxley (in a letter to Romanes) writing 'It is quite conceivable that every species tends to produce varieties of a limited number and kind, and the effect of natural selection is to favour the development of some of these, while it opposes the development of others.'

Why should the pattern of variation be limited in this way? The short answer is that we do not know. Most of us suppose that it has something to do with development. Let me give one example, for which the explanation is fairly clear. Every child can recognize a palm tree, with its long unbranched trunk topped by a plume of enormous leaves. There are many different kinds of palms, all belonging to the monocotyledonous subdivision of the flowering plants, and (almost) all have this same basic appearance. Why? The answer seems to be that the monocots have never invented the process of secondary thickening, whereby the trunk of a tree grows by adding a new ring of wood every year. In consequence a palm trunk, once produced, cannot get thicker, and it cannot produce a twig which will grow to become a branch. (Honesty, and the fear that this may be read by a botanist, forces me to add that a few palm trees, like the cabbage palm, do manage to grow branches; I do not know how they do it.)

More often, we can deduce that there is some constraint on the pattern of variation, without knowing why. For example, monocotyledonous plants are restricted not only by the absence of secondary thickening, but also in the structure of their flowers. These are always constructed in a pattern of threes – three petals, three sepals, and so on. Presumably this has something to do with the way the flower develops, but I do not know what. Interestingly, this tri-radial structure does not prevent the flower becoming bilaterally symmetrical in appearance if this is selectively advantageous: orchids appear to be bilaterally symmetrical because bees are, although their fundamental structure is tri-radial.

Evolution, then, is the result of natural selection acting on a range of variation which is restricted by the process of development. Different biologists may put the main emphasis on one or other of these aspects of the process. The choice is not wholly arbitrary. Thus if the wings of birds were constrained by development to be either triangular or square, then natural selection would probably ensure that vultures had square wings and falcons triangular ones, but the main determinant of wing shape would be development and not selection. If wings can develop any shape, then selection would be the sole determinant. The truth lies somewhere between these extremes.

It is characteristic of the living world that organisms can be classified into groups having a common structure. This classification is hierarchical, so that individuals can be grouped into species, species into genera, genera into families, and so on. I want to concentrate on the more inclusive categories, classes (e.g. mammals) and phyla (e.g. vertebrates). It is characteristic of such categories that the member individuals retain a common structural pattern, despite living in quite different ways. For example, all vertebrates have a stiff backbone, segmented body muscles, a tail extending behind the anus, and two pairs of fins or limbs. (You may quite properly object that human beings do not have a post-anal tail and that snakes do not have legs; however, a comparative anatomist would argue that the absence of particular structures in some members of a group does not invalidate the idea of a common structural pattern.) This common structure has persisted despite the fact that some vertebrates swim, some fly, some climb, some burrow and some run. Why should this be so?

There are, I think, two fundamentally different answers a biologist might give to this question. The first is that there are only certain possible stable states of living matter. The second is that these basic patterns evolved originally as adaptations to some specific way of life, and have been retained despite subsequent changes in ecology.

The first of these two answers is characteristic of the physical sciences. If you asked a chemist why there are only some 90-odd chemical elements, he would reply that these represent the only stable configurations, given the laws of physics and the properties of the fundamental particles. He would not argue that oxygen exists because it is peculiarly well adapted to combine with hydrogen to form water, or that iron exists because it can be magnetized. I have no doubt that the chemist would be right. Those biologists who argue that there are 'laws of form' are committed, it seems to me, to an analogous view: the reason why, for example, there are no animals with bony internal skeletons and six legs is that there are laws of form that prohibit such a structure, just as there are laws of physics that prohibit an element with 11 electrons in its outer shell. The task, then, is to discover those laws.

This is well illustrated by the body plan of vertebrates. The following features are common to the body plan of all vertebrates: a stiff axial rod (the notocord), or its replacement, the vertebral column; segmented body muscles; two pairs of fins or legs; a tail that extends beyond the anus. A supporter of the idea that there are 'laws of form' would, I imagine, hold that these features resemble the structural features of the oxygen atom, and represent one of the relatively few stable configurations possible to living matter: if laws of form do not mean this, I do not know what they mean. In fact, I suspect that the features I have listed exist because our ancestors once lived by swimming by lateral undulations. The backbone and segmented muscles were needed to generate the undulations; the tail was needed to generate a propulsive force; the paired fins were needed to control their movements in a vertical plane, and the number of pairs needed to produce a vertical force through any point is two. For the same reason aeroplanes have two horizontal surfaces, typically a wing and tail-plane.

If I am right, the body plan of vertebrates arose as an adaptation to a particular way of life in an ancestor. It has nothing to do with any law of form. There are many features of our anatomy that are obviously historical accidents. The base of our spine curves because our ancestors were quadrupeds and held the spine horizontally. The air passage from our nose to our lungs crosses the food passage from our mouths to our stomachs because fish had an organ of smell that lay above their mouths. The males among us adopt the curious economy of using the same orifice to pass urine and gametes to the exterior because of a series of historical accidents it would take half a lecture to describe.

Of course, to say that we have backbones because our ancestors swam by lateral undulations leaves unanswered the question of why we, who no longer swim by lateral undulations, have retained this basic plan. A complete answer to this question would be an important contribution to evolution theory. The best I can do is to suggest that this conservatism of basic plans in evolution has arisen because of two other features in the way evolution has to work. The first is that evolution has to occur in small steps, and the second that each step has to improve adaptation, or at least not weaken it. Let me comment on these two points in turn.

First, to say that evolution proceeds in small steps is merely to say that children are rather like their parents. What then of the suggestion that 'macromutations' and 'hopeful monsters' – as Richard Goldschmidt once described them in the 1930s – precursor forms that await a favourable environmental change in order to flourish, have been important in evolution? By a macromutation is meant a single genetic change (that is, a change in DNA structure) that has a large effect on the organism. If such a mutation causes anatomical changes that provide the starting point of a new evolutionary departure, then an individual carrying the mutation has been called a 'hopeful monster'.

A visit to any genetics laboratory will persuade you that there are mutations of large effect. But I think these are all either individuals that lack some structure present in their parents, or have a different number of structures, or have a structure of one kind, say a leg, present in a position where their parents had a structure of a different kind, say an antenna, or, more generally, that the mutation differs from its parents in some way which could be brought about by a small change in the genes controlling development. What we do not find are mutants with new complex structures, not present in any ancestor, and able to perform a complex function: a mutant fly may have a leg where it should have an antenna, but a mutant worm will not have a leg.

The implication of this is that complex new structures are likely to arise in a number of steps, and not all at once. A further constraint arises because, since evolution occurs by natural selection, each step must be an improvement as far as survival and reproduction is concerned, or at least not a serious deterioration. The conservatism of the basic plan, then, is a consequence of the fact that evolution occurs in small steps, by tinkering with what is already there, and not scrapping the existing structure and starting again, as a human designer might do (but, of course, usually does not).

I do not regard this as a complete explanation of the conservatism of development. Such an explanation will require a better understanding of developmental mechanics. For example, it is a curious fact that all vertebrate embryos have a 'notochord' – a stiff rod down the back. In ancestral chordates this rod persisted into the adult, and played an essential role in swimming, but in higher vertebrates it is replaced by the backbone. Why, then, does it persist in the embryo? A possible reason is that it plays a crucial causal role in early development: the cells that form the notocord cause, by contact, the epidermal cells that overlie them to roll up and form the central nervous system. I am not sure that this is an adequate explanation for the persistence of the notochord, but it is probably along the right lines. When we understand more about the mechanisms of development, we may see more clearly why certain features are retained in evolution.

I must return, then, to the question in my title – is Darwinism enough? In a sentence, I think that Darwin's theory of evolution by natural selection is correct, and that it is an essential component of any explanation of evolution. But I do not think it is 'enough' in the sense of being all that we need to know. I have discussed at some length the way in which it is incomplete. It does not explain the processes of development. Since the kinds of varieties that can arise in a given species depend on development, and since the course of evolution is constrained by the variations that can arise, it is obvious that Darwinism is not all that we need to know. My fear is that when people argue that Darwinism is not enough, it is not the absence of a theory of development, or of ecology, that they are worried about. Often I suspect that they are hankering after some kind of Lamarckian inheritance of acquired characters, or some Teilhardian inner urge towards the omega point. If so, they would be better to stick with Darwin.

5

Is Biology an Historical Science?

Brian Goodwin

History vs Logic

Different sciences are distinguished not only by their subject-matter, but by the way problems are approached, their analytical style. What characterizes biology is that virtually all questions are understood to be essentially historical ones, relating to evolution. No matter what aspect of the structure or behaviour of organisms you may be studying, what ultimately defines how that enquiry makes biological sense is the way it contributes to an understanding of the origin, the persistence, or the decline of species. As Dobzhansky (1973) put it in the title of one of his articles, 'Nothing in biology makes sense except in the light of evolution.' This historical approach to biological problems was, of course, due primarily to Darwin. When combined with an appropriate theory of inheritance as provided by Mendel and Weismann, it gave biology a conceptual framework of such versatility that it has kept biologists busily engaged in extremely productive research for over a century. But things are now inexorably changing. To see why, let us look at what it means for a science to be based on historical explanations.

Let me consider first a non-biological example, because the message comes across more clearly when it is in an unfamiliar context. Suppose you asked somebody the question:'Why does the earth go round the sun in an elliptical orbit?' and you got the answer: 'The earth goes round the sun in an elliptical orbit this year because it went round the sun in an elliptical orbit last year, and nothing has happened to change it.' Would you be satisfied with this? It is an historical answer, in the style of biological explanations. It also happens to be a perfectly correct, though limited, statement. The reason why it is not satisfactory is that, to use technical language, it describes only necessary conditions for the earth's elliptical motion (position and velocity of the planet at some arbitrary point in the annual cycle, taken to define its beginning, say the New Year). What is missing is any reference to Newton's inverse square law of gravitational attraction, which is what ultimately determines that one of the possible forms of motion of a body such as the earth attracted to a massive body such as the sun, is an ellipse. This law, together with the particular history of the planetary system, determines the *sufficient* conditions for the earth's elliptical orbit, and it is this type of complete explanation that physicists seek. Biologists, on the other hand, settle for incomplete, historical explanations, based only on necessary conditions, though in the other sciences this is normally used only as a last resort.

Let us turn now to a typical biological example. Suppose you ask a biologist 'Why do my arms and legs have this form?' The answer you are most likely to get is: 'Your five-digit (penta-dactyl) limbs are inherited from a common ancestor of the tetrapods, believed to be a primitive amphibian with a basically similar five-digit limb, but there are modifications in the bones of human limbs for functional (adaptive) reasons.' This is clearly an historical answer, based upon inheritance and natural selection. It explains *form* by assuming an historically given form in an amphibian 'common ancestor', who is unknown (and, in fact, unknowable), together with postulated selection forces, also unknown, which result in modifications of the ancestral structure. The approach has resulted in numerous conceptual difficulties, many of which can be found in a fascinating paper by the eminent biologist, Sir Gavin de Beer, entitled *Homology: An Unsolved Problem* (1971). As he points out, the whole science of comparative anatomy (hence of taxonomy) is based upon the recognition of homologous (structurally similar) forms in different groups of organisms; and distinguishing these from analogous (functionally similar) forms. Thus it is acknowledged that structural similarity, such as the patterns of the bones in the wings of a bat and in the fore-limbs of a horse, involve a much deeper and more significant notion of affinity than the functional similarity of the bat's wing with the wing of an insect. This had been established by the pre-Darwinian comparative anatomists, who created the science of rational morphology, in which the forms of organisms were related in terms of the concept of homology, understood as equivalence under trans-formation. Thus the bone pattern in the wing of a bat and the fore-limb of a horse can be transformed into one another by modifications in the sizes and number of bones, their relative positions remaining unchanged; whereas the structural relationships between the wings of a bat and those of an insect are much more remote, there being no equivalent elements in the two structures that can undergo transformation. The science of form resulting from the rational morphologists, such as Goethe, Cuvier, Geoffrey St Hilaire, Reichert, and Owen in the late eighteenth and early nineteenth centuries employed the concept of homology in a purely logical or abstract sense. It later led to the celebrated work of D'Arcy Thompson, who demonstrated homological relationships in terms of mathe-matical transformations which change one organismic form into another. This is the systematic procedure whereby structural similarities are described and classified. With it, one could hope eventually to develop a rational taxonomy of organisms, a biological classification scheme which describes the structural affinities between different transformations. An equivalent of this in physics is the rules of transmutation of the elements, defining how oxygen, say, can be transformed into nitrogen by a loss of electrons, protons and neutrons, and a spatial reorganization of the remaining elementary particles, which continue to obey certain relational rules (as do the bones in a limb). This was the goal of rational morphology, to render the whole of the biological realm intelligible in terms of general structural principles or 'laws of form'. As we shall see, their research programme had its limitations, one of which is that such a goal cannot be achieved simply by comparisons of adult morphologies and requires a science of how the forms are generated. But it did clarify and define a centrally important biological problem and a systematic approach to it, seeking to make biology as a whole intelligible in terms of the intrinsic, logical properties of organisms and their parts.

However, Darwin profoundly altered the meaning of homology when he argued that it should be understood in terms of descent from a common ancestor, replacing a logical definition by an historical one. This imposed a serious limitation on the concept, since a logical definition of the structural relations of real entities may or may not describe also an historical relationship, whereas an historical definition restricts relationships to those of contingent material continuity. To make this point clearer, let us consider one of the difficulties that de Beer gets into in using Darwin's definition of homology. William Bateson, in his classic study of biological forms and their variations (*Materials for the Study of Variation*, 1894), used the term 'serial homology' to describe the structural similarities between different parts of the same organism, such as the different body segments in insects, or the fore- and hind-limbs of tetrapods (four-limbed vertebrates). Bateson did not subscribe to Darwin's definition of homology, and in fact he was very critical of the whole Darwinian approach to biological form and its evolution because of its excessively historical and atomistic emphasis (i.e. its focus on variation of individual parts of organisms rather than transformations of whole patterns, which Bateson emphasized and for which he gave good evidence). De Beer correctly points out that, in terms of Darwin's definition, serial homology such as the structural relations between the fore- and hind-limbs of a particular species, 'is not real homology, as fore-limb and hind-limb cannot be traced back to any ancestor with a single pair of limbs.' This view thus leads to the conclusion that our arms and the wings of a bat or the fore-limbs of a horse are homologous (structurally similar), but that our arms and legs are not. This violates our common-sense notions of similarity.

Clearly, a logical definition of homology, based upon the intrinsic spatial relationships of structures, imposes no such constraint, and it is the one that the phenomena demand. This is particularly clear in the case of insect segmental structures and their transformations. The appendages associated with particular segments, such as eyes and antennae, legs and wings, are all transformable one into the other by a class of genetic mutations called homoeotic. So they are equivalent under actual transformations caused by gene mutation. It so happens that they are mutually transformable under environmental disturbance also: if *Drosophila* (fruitfly) embryos are exposed to stimuli such as transient temperature change or chemical treatment (ether, boric acid, nicotinic acid, etc.) at particular times during their development, segments produce legs instead of antennae or wings instead of balancing organs (halteres), or other transformations, just as they do when the organisms have particular mutant genes. This evidence requires that we recognize insect segments as homologous in the sense of the rational morphologists (equivalent under transformation) rather than in the Darwinian sense (descent from an insect ancestor with a single segment, which makes no biological sense). As pointed out earlier, this does not exclude the possibility of historical continuity of certain structures in organisms, but it does not limit us to such continuity, and so avoids the conceptual difficulties that de Beer encountered.

Genes and morphology

It was thought that a way out of these difficulties, while retaining a historical perspective on the study of organismic form, would come from the analysis of the genetic basis of structural similarity. Thus if it could be shown that homologous structures involve the action of the same genes, then a causal genetic analysis, with historical continuity of these hereditary elements, would be possible. However, de Beer points out that a mutation in a single gene can cause morphological changes in characters that are not homologous, so that one cannot say that characters controlled by identical genes are necessarily homologous. Furthermore, identity of morphology of particular parts of different individuals does not imply similarity of genotypes, as de Beer describes in relation to the eyeless mutation in *Drosophila*. This mutant gene results in a failure of eyes to form, but normal eye morphology returns (after a certain number of generations) as a result of changes in other genes, despite the continuous presence of the eyeless gene.

Thus the same structure in *Drosophila* can occur through the action of different genes. It is even the case that the same type of appendage (e.g. legs) can be produced in one individual by the action of different genes, as shown by Morata and Kerridge (1982). In this case, legs of normal morphology are produced in place of antennae in a particular mutant (called antennapaedia), but different genes are involved in the formation of these legs compared with normal ones. In the unicellular organism *Tetrahymena* the same detailed morphology occurs in different species despite great differences in the genes (and hence the proteins) involved in producing this morphology (Williams, 1984). Thus particular genes are neither necessary nor sufficient for particular morphologies, so that there is no way of identifying historical constraints on form with constraints on particular genes. This is not to say that there is no relationship between genes and morphology, because there are many examples in which gene mutation results in changed structure. However, there are many different combinations of genes that are compatible with a given morphology; and conversely, with a fixed genotype many different forms can be produced, even in a constant environment. The latter conclusion comes from a variety of studies, such as those showing that the same proteins can be assembled into structures, such as flagellae, with different morphologies (Oosawa et al., 1966); and that individual organisms, all of whose cells have the same genes, can have different appendages on different sides of the body (e.g. *Drosophila* homoeotic mutants with a leg instead of an antenna on one side of the body, but a normal antenna on the other; smooth newts with one limb that has palmate newt morphology, webbed fingers instead of separate, etc. (Roberts and Verrell, 1984)).

De Beer concludes his paper with the question whether homologous structures are controlled by non-DNA mechanisms, and one is virtually driven to an affirmative answer by the evidence, although it is necessary to be very careful in describing exactly what is meant by this. One thing, however, is clear: the historical approach to the analysis of biological form, whether at the level of gross morphology or in terms of genetic mechanisms, has failed and we must approach the problem from a different perspective. I argued above that rational morphology offers a more satisfactory definition of the key concept of homology, but that the approach to an understanding of the structural relationships of organisms through the study of adult morphology has some serious limitations. This is because structurally similar adult forms may have been arrived at by different developmental routes. For example, the five-fold radial symmetry of starfish suggests a structural affinity with organisms such as hydroids or jellyfish, which also have radial symmetry of adult body plan. However, the adult starfish symmetry arises by loss of bilateral symmetry in a larval form. So we need to know the whole of the generative process that an organism undergoes to describe adequately its place in the scheme of biological process. This scheme therefore involves a dimension other than adult morphology, which is the dimension of ontogenesis, or reproduction.

The nature of the reproductive process

There is an enormous variety of patterns of reproduction which organisms undergo, but they all have a basic feature in common: from a part of an organism a whole is produced. The part may be a single cell that develops into a whole multicellular organism by a complex series of stages, such as occurs in sexual reproduction from an egg; the part may be an aggregate of cells, as in asexual reproduction by budding; or the part may be a single cell which gives rise to two single cells, as in the reproduction of unicellular organisms such as bacteria, in which case the part is equal to its upper limit, which is the whole organism.

The process of parts giving rise to wholes also occurs in regeneration, as in the formation of a whole, normal limb from a stump in newts or salamanders; or in the reconstitution of a whole organism from almost any fragment in hydroids and planarian worms. Reproduction and regeneration (more generally, healing, making whole) are extremely basic properties of organisms, giving the biological domain those properties of continuity and stability over individual lifetimes and between generations which are the distinctive features of the living realm. It is on this foundation that evolution, the historical expression of the potential of the living state, is based. Inheritance involves both the transmission of distinctive organizational features of living organisms from generation to generation, and the transmission of the specific characteristics which distinguish one organismic type from another. This is not the commonly accepted definition of inheritance. Most biologists consider that the term inheritance should be restricted in meaning to the transmission of inherited *differences* between organisms, since in their view evolution is about the origin of different species. This definition suits the discipline of genetics, since the analysis of what makes organisms different from one another is what this subject is all about, and population biology is based upon genetics.

However, if our objective is to understand the structure of the biological domain as a whole, then it is necessary to understand both what it is that makes organic forms similar to one another and what makes them different. The fact that organisms can have very different genotypes but exhibit the same morphology, as in the different species of *Tetrahymena* mentioned above, that look identical but have different structural proteins, means that we cannot seek an explanation of this morphological similarity in terms of similar genes. What is similar is the morphogenetic process and the forces operating that generate the form; and this similarity can occur despite structural gene differences. In a unicellular organism such as *Tetrahymena*, the morphogenetic process involves both short-range forces (Ångstrom dimensions) and long-range order (micrometres, μm), as in multicellular organisms. The former are reasonably well understood, since they underlie basic molecular processes such as enzyme action, antigen–antibody interactions, and the base-pairing forces involved in the self-replication of DNA; i.e. these are the intermolecular forces on which molecular biology is based. Long-range order shows up in *Tetrahymena* as the force which determines the relative positions of the oral apparatus (mouth) and the contractile vacuole pore (excretory apparatus), separated by some 30 μm, and the positions of these structures when they are produced anew during cell division. In the metazoa they underlie the morphogenetic movements which are co-ordinated over distances up to several millimetres in extent. These forces remain to be adequately explained, although there are plausible candidates in the viscoelastic forces associated with the properties and behaviour of the cytoskeleton and the cytomusculature, a scaffolding of contractile and mechanical filaments that exists in every eukaryotic cell, extending throughout its interior and making mechanical connections with the membrane.

The morphogenetic properties of this system have only recently begun to be systematically explored (Odell et al., 1981; Oster et al., 1983; Goodwin and Trainor, 1985). As argued elsewhere (Goodwin, 1984), this approach to biological form and its transformations through a study of the forces that are operating across membranes and through the cytoskeleton, associated with electrical potentials and ion currents – particularly the effects of calcium and pH on the mechanical and contractile state of the cellular scaffold – has already given important insights into the nature of the processes that underlie morphogenesis and reproduction. The proposition that I am presenting here is the somewhat old-fashioned view that organisms may be regarded as a particular state of organization of matter, 'living substance' if you like, and that an understanding of this organization will provide insights into those distinctive features which we associate with the biological realm, such as reproduction and evolution. These require an understanding of morphological stability and change which are not reducible to stability and change in either genomes or in the environment, since organismic morphology can remain unchanged despite changes in both of these sets of variables. This shows us that the dynamics of reproduction and evolution do not reduce to a description of internal (genetic) and external (environmental) forces that push organisms around as passive mechanisms. Organisms constitute a third, integrative level of order in organic process, turning random internal and external changes into organized transformation (which includes invariance). Clearly the problem that faces us is that of identifying the appropriate causal factors for understanding morphological relationships between organisms, both as they have emerged in time (the problem of evolution) and as they relate logically (the problem of taxonomy, classification). We have seen that an historical approach to the latter problem leads to conceptual contradictions; and that the former problem cannot be analyzed in terms of genes. A worthwhile proposition to pursue is that the answers to these problems may be in the nature of the organizational forces which are intrinsic to the living state and provide it with the basic dynamic that results in biological process, developmental and evolutionary. We have seen that a basic characteristic of the reproductive process, on which evolution depends, is the capacity of a part to give rise to a whole. Such spatiotemporal organizations of matter are known as fields, and their properties are described by specific types of equation, expressing the operation of particular types of forces. Let us now consider the characteristics and the behaviour of simple examples of fields, to see if they give us some insight into the type of organization needed to understand the biological process.

Organisms and fields

A very simple example of a physical field is the velocity field that describes the movement of a liquid under particular conditions of motion. For example, if a bath-tub is filled with water and then the plug is removed, the water flows down the plug-hole with spiral motion, left- or right-handed depending upon the local motion of the liquid caused by the removal of the plug (only under very special conditions is it determined by the Coriolis force, due to the earth's rotation, resulting in different-handedness in the northern and southern hemi- spheres). Right- and left-handed spiral flow are the two stable forms of motion available to this state of matter under these conditions, and these forms or morphologies are the stable solutions of the Navier–Stokes field equations which describe this state of organization of matter. It is common experience that these liquid morphologies can be transformed one into the other. Thus, suppose you have right-handed spiral flow out of a bath-tub; then you can convert it into left-handed spiral motion simply by swirling the water around in a left- handed direction in the neighbourhood of the hole, and a stable left-handed morphology will be produced. Now, if you were asked the question: 'What caused the left-handed spiral flow?' you might be tempted to answer: 'The motion of my hand.' This would be a correct, but incomplete, response, just as the response to the question about the earth's elliptical orbit around the sun in terms of last year's orbit was correct but incomplete. Just as a complete answer in the case of the earth's motion requires a statement about the gravitational field as defined by Newton's inverse square law, which is what makes elliptical motion a possibility, so in the case of the bath-water it is necessary to define the properties of the liquid state in terms of the Navier–Stokes field equations, which properties make spiral flow a possibility. Your hand simply selected or stabilized one of the possible morphologies available to that state of matter under those conditions (a container with a hole, with gravity acting upon the liquid). Using this example as a metaphor for the biological point I am making, one could say that a geneticist, being interested in what makes one form different to another, focuses on the biological analogue of the force that turns a right-handed spiral into a left-handed spiral. For this type of phenomenon occurs in biology. An example is found in the snail, *Limnaea*, in which the normal or wild-type has a right-handed spiral shell; but a mutant gene can cause left-handed coiling. Geneticists then tend to conclude that the mutant gene causes the left-handed form. But this is 'cause', I suggest, only in the sense that the motion of your hand can cause a left-handed spiral in the bath-water. It is an incomplete explanation, for it leaves out an account of what makes a spiral form possible in the first place, either in the water or in the snail. So I am arguing that, just as liquids are states of matter which are organized in such a way that certain forms or morphologies are possible (spirals, waves, jets, etc., under various specific conditions), so organisms are states of matter organized in such a way that particular forms or morphologies are possible. And just as liquids can have different molecular compositions (water, alcohol, benzene, etc.) and yet show the same morphology (e.g. spiral flow), so organisms can have different molecular compositions (different genes and gene products) and show the same morphology, as in *Tetrahymena*. However, it is also possible to alter the morphology of liquid flow by changing its composition. For example, if a polymer is added to a liquid so that its viscosity is altered, then there will come a point where flow down a plug-hole ceases to be spiral and becomes straight; and if the polymer sets in the liquid, converting it to a solid, then mass flow ceases and different equations are required to describe a different state of matter, including stress–strain relations and elastic waves, for example. So composition can certainly affect morphology and behaviour, as it does in organisms; but the relationship is not simple and direct, as is implied in the genetic programme model of embryonic development, wherein gene activity is described as the 'cause' of morphogenesis. Gene activity can influence morphology, but need not. It is certainly not a sufficient explanation of biological form and its transformations.

The analogies used above between organisms and liquids are obviously so crude as to be misleading if carried any further than the simplest idea that organisms reveal a state of organization of matter, the living state, which is capable of taking on certain forms. This implies that there are morphologies that are unavailable to organisms; more generally, it implies that organisms are natural kinds or forms, just as spiral flow is a natural form in liquids, so that there is a rational taxonomy of organismic morphologies. However, we will not know what these forms are until we have an appropriate field theory (and field equations) that describes how the forms are generated, as the Navier–Stokes equation does for liquids. Just to add a note of caution to the discussion, it should be realized that new solutions of this equation are still being discovered, many years after it was derived. And we have not even derived the appropriate field equation to describe the living state despite important progress in this direction. However, none of this alters the logic of the proposed research programme. Furthermore, the combination of subtle analytical (mathematical) tools and enormously sophisticated computational and graphical procedures may well make it possible to explore a significant part of biological 'morphological space' before very long, and to discover the basic structural patterns in it. Such a research programme, combined with experimental methods which reveal both the phenomenology of morphogenesis and the relations between composition and form in organisms, is more than a possibility. It has now become a necessity, due to the breakdown of other approaches to this problem. What could emerge is an adequate conceptual and theoretical understanding of the organizational principles of the living state as a self-transforming field with characteristic dynamics.

Natural selection and the emergence of novelty

In the discussion so far, nothing has been said about natural selection, which is generally regarded as one of the primary forces responsible for determining the morphologies of existing and extinct organisms. This view arises from the proposition that organisms must be adapted to particular ways of life, and this adaptation includes morphology as a component: the long limbs of horses enable them to escape predators in the open grasslands that are their natural habitat; bats' wings allow them to exploit an aerial niche unavailable to other mammals, etc. This focus on adaptation is the essence of Darwinism, and there is no doubt that some notion of stability of organismic lifestyles is absolutely essential to any evolutionary theory. However, it is necessary to keep the horse before the cart. Selection obviously cannot act upon morphologies that do not exist. The generative process in any particular species or genus will define some set of morphologies as the potential set for evolution. Exactly how limited this set is could be a result of either selection itself, restricting the range of variability through a canalizing mechanism; or it could be due to the intrinsic dynamics of the developmental process. The latter may itself have different degrees of variation. On the one hand, there are developmental processes that impose absolute constraints on possible morphologies, as appears to be the case in *Drosophila* with respect to bilateral symmetry: no-one has succeeded in selecting a morphologically asymmetric strain of these flies. On the other hand, there are processes which result in great heritable variability in morphology, such as the exact positions of bracts (bristle-like structures) in *Drosophila*.

There is, however, another dimension to developmental dynamics which is of great evolutionary importance, and that is the emergence of new developmental processes that give rise to novel morphologies. For example, tetrapod limbs, such as those of newts, crocodiles, birds, and mammals, could not be generated until a certain combination of processes arose in vertebrate embryos. Selection certainly did not produce these in any direct sense, since until the structures were actually produced there was nothing for selection to act on. So how did they arise? They must have originated from a new combination of developmental processes. Whether this arose from random variation of genes affecting the dynamics, from indirect selection (effects on one aspect of development due to correlation with other aspects which are selected), or from a direct influence of the environment on the developing organism which persists via maternal inheritance and produces a dynamic bias which is eventually stabilized by genetic changes, are questions for empirical study. We do not understand much about such processes at the moment. It may be that there are universal properties of the living state that strongly constrain the set of such new possibilities, so that they may be predictable in the sense of being theoretically describable. On the other hand, it may turn out that the living state is compatible with an enormously varied and virtually continuous set of possible morphologies, so that there are practically no intrinsic discontuities between the forms that *could* be produced (in contrast to those that *are* produced). The latter is basically the Darwinian view, so that taxonomy is essentially genealogy, history, what selection has winnowed out for survival from the continuum; the former is the view that organismic morphologies are separated naturally by discontinuities, so that there is a rational taxonomy and organisms are natural kinds. It could be that the truth turns out to be somewhere between these positions, some aspects of organismic relationships arising from the intrinsic properties of developmental processes and some from historical factors. On the other hand, it seems to me more likely that an adequate resolution of the conflict will actually carry us forward to a quite different perspective regarding the nature of the causal processes operating in organic evolution. Rather than some compromise between the outside view espoused by Darwin that emphasizes adaptation and natural selection, and the inside view that stresses the intrinsic properties of biological organization, we need a conceptual framework that transcends these dualisms and provides a basis for understanding transformation and stability in continuous process.

6

Sociobiology: the New Storm in a Teacup

Richard Dawkins

One of the more irritating things that has happened in the sociology of biologists in recent years is the invention, apparently for political purposes, of an alleged 'debate' or 'controversy' over sociobiology. Sociobiology was originally defined 'as the systematic study of the biological basis of all social behavior' (Wilson, 1975, p. 4). It is a field of study, then, not a point of view. As such it is no more controversial, in itself, than metallurgy or ornithology. Of course you may disagree with the views of individual sociobiologists, and individual sociobiologists may disagree with one another just as, I dare say, individual metallurgists disagree with each other. But you can no more disagree (or agree) with socio-biology as a whole than you can disagree or agree with metallurgy as a whole.

Now the reader may object that there is something dis-ingenuous in my first paragraph. However neutral the original definition of sociobiology may have been you cannot, it will be said, deny that it has become identified with a stance, a point of view. By the ordinary processes of linguistic evolution (aided, in my opinion, by some systematic misrepresentation), 'sociobiology' has come to assume a new and far from neutral meaning just as, at one time in America, the originally neutral phrase 'Law and Order' became a code symbol for non-neutral racialism. Incidentally, the same kind of thing has happened to 'ecology' and to some extent to 'sociology'. If you call somebody an ecologist, how shall we know whether he or she is a student of organisms in relation to their environments, a 'green' politician, or both?

Very well then, but if you insist on using 'sociobiology' as a label for a set of opinions rather than for a neutral field of study, you must then check up, before labelling a particular individual a sociobiologist in your sense, whether he or she in fact holds those opinions, or is really a sociobiologist only in his or her own sense. If you do not do this person the courtesy of checking up, you will be in the same position as somebody who assumes that a professor of sociology must, by definition, be a left-wing radical.

Now there may be those who question the neutrality of the allegedly neutral definition: 'the biological study of all social behaviour', especially where human social behaviour is concerned. This is because, for them, 'biological' is not a neutral word. In the first place it may imply a certain philosophical stance which they would probably characterize as 'reductionist' and 'determinist'. I shall deal with these words later. In the second place they may see something demeaning to 'human dignity' in studying human social behaviour biologically, because it implies that we are 'mere animals'.

Unfortunately – for it is very widespread – this second objection to the 'biological' study of human behaviour is based upon an anachronistic taxonomic misunderstanding. There is no sensible usage of the word 'animals' that places *Homo sapiens* outside the category and all other species in it. Obviously I cannot stop you classifying organisms in any way you choose: you could divide them into 'Pricklies' (hedgehogs, porcupines, spiny anteaters, porcupine-fish and sea urchins) versus 'Animals' (all the rest, including humans), but it would not be a sensible or a useful classification. Actually, to be fair, this would be even less sensible than the 'humans versus animals' classification, because 'Pricklies' is itself a heterogeneous category. A closer analogy to the 'humans versus animals' classification, and one that cuts nearer the bone, would be a Nazi classification into 'Aryans versus animals'.

All species are twigs at the ends of branches on the bushy tree of evolution. It is a tree in the mathematical sense of a branching pattern, rather than in the sense of something that has height like a ladder. There are no 'higher' or 'lower' animals, there are only main branches and subsidiary branches and tiny twiggy branches. This is equivalent to saying that animals have very distant cousins, distant cousins, closer cousins and very close cousins. The details of the subdivisions of the tree can be argued about, but there can be no doubt of our general position. We are not bacteria, we are not plants and we are not fungi. We firmly belong with the large group which, for want for a better word, we have to call animals. Within animals we are vertebrates, within vertebrates we are mammals, within mammals we are primates, and within primates we are – by exactly the same logic – apes.

We are closer cousins to some apes (chimpanzees and gorillas) than those apes are to other apes (orang utans and gibbons). Therefore if we are going to use the word apes at all, the word cannot be used to exclude ourselves and include both chimps and orang utans. There is a tiny twig labelled apes buried somewhere in the dense thicket of branches on one side of the tree of life. This little ape twig divides very near its tip into a cluster of even tinier twigs, one of which subdivides further to form a twiglet labelled *Homo*. The conventional division into humans *versus* 'animals' is a preposterous distortion of the tree of life. Apart from anything else, it is unpleasantly patronizing to chimpanzees to lump them with jellyfish, tapeworms and amoebae in a category that does not include humans.

Now when I have put this point to audiences, it has provoked a certain amount of agitated jumping up and down: 'animals' can't reason; 'animals' can't talk; 'animals' can't read, do science, play music. Apart from the fact that this begs the question (the animals called humans can do all these things), and some non-human species certainly can reason, the best answer to this kind of argument is 'So what?' Many animal species have their own peculiarities which, however remarkable they may be, do not entitle them to be excluded from the animal kingdom. Swifts can fly hundreds of thousands of miles without ever alighting, but we do not, on this account, find it necessary to erect a dichotomy between 'swifts' and 'animals'. Swifts are animals, albeit animals with remarkable powers, and it is as biologists that we study those powers. Similarly humans have remarkable powers, but this does not mean that they are not animals, and it does not mean that we should not study the biology of those powers. To say that humans are not 'animals' because they can talk, do mathematics, etc., is like saying that the Concorde cannot be an airliner because it is supersonic. Humans are animals and, as such, part of the legitimate field of study of biologists.

This is all very well, it may be said, but the logical upshot seems to be the extreme conclusion that not only sociology and anthropology but history, literature, art and all the subjects known as humanities should be regarded as sub-departments of biology. In one sense this is true, but it should not be feared as a take-over bid. It does not mean that the Departments of History, Art and the rest should all be housed in the Biology Building of the university. Biology in its turn could be called a branch of physics, but it is such a large and peculiar branch of physics that it still makes sense for biology to be studied in its own right, and by people untrained in the matters that preoccupy conventional physicists. Similarly the study of human history may be, in an extreme technical sense, a branch of biology, but it is still such a large subject that it can, and should, be studied in its own right, and by people untrained in (other aspects of) biology. It is not that I am objecting to the category 'human', as a label for a legitimate area of study in its own right. What I am objecting to is the category 'animal' where it is taken to *exclude* one particular species, whether our own species or any other.

Human social behaviour is undoubtedly a legitimate candidate for biological study even if, when we study it as biologists, it eventually turns out that it is *in fact* so far removed from the principles that govern the behaviour of other animals that present-day biological expertise can make no useful contribution. If biologists fail to understand human social behaviour, it will not be because they are obviously doomed to failure as a matter of definition: 'because humans just are not animals, and that is all there is to it'. No, if we fail it will be because we have not got to grips with the unusual biological complications presented by the peculiarities of this particular species of animals, for instance language.

We should distinguish three ways in which biologists might approach human social behaviour. We might look at the facts of the social behaviour of particular animal species, say three-spined sticklebacks and grey-lag geese, and then make an inductive generalization to humans. This is the approach for which it was once fashionable to criticize Konrad Lorenz and Desmond Morris. My only comment at this stage is that it is in principle just as objectionable to generalize from sticklebacks to geese as it is to generalize from sticklebacks to humans. The second thing the biologist might do – this is the approach recommended by Niko Tinbergen (e.g. 1968) – is to set aside the facts of stickleback and goose ethology and, instead, apply to humans the same *methods* as were used to study sticklebacks and geese. For Tinbergen this largely means methods of systematic and detached observation. I shall not comment on this either, except to remark that in practice those ethologists who have followed Tinbergen's advice have concentrated on *non-verbal* behaviour, and it is important to realize that this should be for practical reasons only, not as a matter of principle. The third thing the biologist might do, and the thing that interests me most, is to apply to humans not the detailed facts of other species' behaviour, nor the observational methods developed for studying other species, but the *principles* that apply to all living things, regardless of the details of particular applications to particular species. I especially mean Darwinian evolutionary principles.

In a sense this is an aspect of the biological approach that is not strictly neutral like metallurgy, but part of a committed stance. Biologists do not, as a matter of definition, have to be committed to the Darwinian view of life but, as a matter of fact, all serious modern ones are – give or take a few minor differences of detailed interpretation. The third approach of the biologist to human social life, then, is to say: We have good reason to be committed to the view that the human body and brain are the products of Darwinian evolution. Now, what does this tell us, or lead us to suspect and examine further, about human social behaviour? If I were asked to encapsulate in a phrase the stance, or commitment, of modern socio-biologists, I would say that they are Behavioural Darwinists. So, of course, are ethologists, and incidentally I therefore see no very great need for the word 'sociobiologist' to have been coined at all. Still, we seem to be stuck with it now.

By far the most important thing the Darwinian assumption tells us about ourselves is the answer to the fundamental question, until recently usually answered in religious terms, of why we exist at all. The answer to this fundamental question is that our bodies are mechanisms for the preservation and propagation of our genes. This conclusion will not be doubted by anyone who has seriously considered the matter. Even a vocal opponent of 'sociobiology' has put it thus, in the particular context of the problem of selfishness versus altruism:

> Natural selection dictates that organisms act in their own self-interest. . . . They struggle continuously to increase the representation of their genes at the expense of their fellows. And that, for all its baldness, is all there is to it; we have discovered no higher principle in nature (Gould, 1978, p. 261).

Gould was not speaking of humans in particular, but it would certainly never have occurred to him to exclude any particular species from the generalization. It is not that kind of generalization.

Our bodies, then, are machines for propagating the genes that made them; our brains are the on-board computers of our bodies; and our behaviour is the output of our on-board computers. This is all very well as a general statement, but does it tell us anything, in particular, about human social behaviour? It is quite possible that it does not. It could be that, although the human brain exists in the first place as part of a gene-preserving machine, the conditions under which it now lives have become so distorted that it is no longer helpful to interpret the detailed facts of human social behaviour in Darwinian terms. Indeed, my own view is that it is downright naive to look at the social life around us and try to interpret the actions of individuals directly in terms of survival value or gene preservation. The operative word is 'directly'. I can illustrate this with an analogy.

Moths fly into candle flames and other sources of artificial light. This habit seems to be very harmful to their survival. At worst they get burned, and at best they waste their time apparently 'trapped' in the vicinity of the light. Now suppose we were debating the question whether 'self-immolation behaviour' in moths is built in by natural selection. When put in these terms, the very idea is absurd. How could natural selection possibly have favoured a behaviour pattern whose effects are manifestly harmful to the animal concerned. Some ingenious 'sociobiologist' might dream up a far-fetched explanation: perhaps the moths are sacrificing themselves for the benefit of their kin, removing themselves from this world as would-be competitors for food. But such an explanation would be laughed out of court, and rightly so.

The trouble began with the way in which we chose to characterize the behaviour we were trying to explain. We called it 'self-immolation behaviour', and this is indeed a reasonable description of what we observe. But we could have labelled it in some other way, and then we should have asked our question about natural selection very differently. A moment's reflection tells us that candles, and other sources of artificial light, are not a part of the natural environment in which the vast majority of the evolution of the moths' ancestors has taken place. Right, then, what *was* in the night environment of the moths' ancestors that might be being confused with a candle flame by the moths' sensory apparatus? The obvious answer is the moon or some other celestial body. If Tinbergen's (1953, p. 66) sticklebacks could confuse a red mail-van seen through the window with a red male stickleback, and give the mail-van an aggressive display, surely a moth could confuse a candle or a light bulb with the moon. So let us try labelling the behaviour differently. Instead of calling it 'self-immolation' behaviour we label it 'moon-approaching behaviour'. We now regard candle-approaching as a mistaken byproduct of moon-approaching and we ask, has natural selection favoured moon-approaching behaviour?

Once again, the idea appears daft. Why should a moth want to go to the moon? Not only is the moon an inhospitable place, but it would be energetically much too costly for a moth to reach escape velocity. Once again the problem lies in the way have phrased the question. By labelling the behaviour 'moon-approaching behaviour' we have again loaded the dice against a sensible application of natural selection theory. Our problem is to find a way of labelling the behaviour, a way that has a chance of leading us to a sensible Darwinian hypothesis. So, let us think a little more deeply about how a moth might actually use a star or the moon.

Optically speaking, the moon or a star are at infinity. This means that rays of light from the moon are parallel when they strike the moth. This in turn means that they can be used as a compass, to steer a straight course. There is abundant evidence that insects, from many diverse orders, make heavy use of this technique of navigation. Ants and beetles, for instance, have been convincingly shown to use the sun as a compass. Bees have sophisticated time-compensation mechanisms to cope with the fact that, due to the rotation of the earth, the sun appears to move round the sky. It is entirely plausible that a night-flying insect like a moth would use a moon compass. But why should this lead to their flying into the candle flame?

That insects should evolve sophisticated time-compensation mechanisms to cope with the rotation of the earth makes sense because the earth has always rotated, throughout the long period during which their ancestors were naturally selected. But there is one new feature of the night sky which their ancestral experience has not equipped moths to deal with. This is the presence of glowing bright objects that are not at optical infinity: candles, electric lights and so on. Now, what happens if you try to keep a light source at, say, 30 degrees to starboard, if that light source is not a celestial body at optical infinity, but a candle at one yard? The rays are no longer parallel, they radiate out from the candle. It is easy to show that the moth, if it maintains a fixed (acute) angle to these rays, will describe a neat logarithmic spiral into the candle flame. Remember that the moth probably has no idea why it does it. Its on-board computer is wired up in such a way that it behaves in this way, and in ancestral times when rays could be relied upon to be parallel, this always had the desired effect that the moth flew in a straight line.

Now that we have characterized the moth's behaviour in an appropriate way, not as 'self-immolation behaviour', nor even as 'moon-approaching behaviour', we can begin to apply natural selection theory sensibly. The behaviour is now labelled 'celestial compass navigation behaviour'; it has obvious survival value, and 'self-immolation' is now seen for what it is, an aberrant byproduct which only arises in an unusual environment.

The moral is clear. When we look at animals, including ourselves, and ask Darwinian questions about 'survival value', we can expect to get a sensible answer only if we think about the context in which natural selection acted upon that behaviour, and label the behaviour accordingly. It is no use asking about the survival value for sticklebacks of chasing mail vans. We can ask about the survival value for sticklebacks of chasing red objects, and then ask what red objects might have been present in the environments in which the ancestors of the stickleback were naturally selected. The answer in this case is obvious – the red bellies of other male sticklebacks. But it may not always be so obvious, and it certainly is not usually so obvious in the case of our own species.

That we are living in surroundings that are grossly different from those in which our ancestors' genes were selected is undoubted. What is more interesting is that the relevant changes have been brought about in a progressive, quasi-evolutionary manner over historical time, and are the more profound because of it. To look at some aspect of modern human social behaviour, custom or manners, say a supposed desire among business executives for bigger desks, and to ask 'What is its Darwinian survival value?', is naive in the same kind of way as our initial question about moths was naive, but in the human case it is much more so. If we are to get anywhere at all, the very least we must do is to rephrase the question sensibly. In the moth example we employed the following re-writing rule. We replaced 'self-immolation be-haviour' by 'maintaining a fixed (acute) angle to light rays assumed to come from infinity'; and then the question made sense. Goodness knows how we might have to rewrite the question about executives' desks before we had it in a shape in which a Darwinian might make sense of it. It might never make sense.

The problem of discovering the correct re-writing rules for Darwinian questions is probably more difficult than the problem of answering those questions when once they have been properly re-written. One thing we can say is that, since natural selection can work only upon genetic variation, a major qualification of a good 're-writing' rule is that it should identify some aspect of the behaviour that shows (or showed) genetic variation. This, as we shall see, is why socio-biologists are so careful to phrase their functional hypotheses in terms of genes: a carefulness which, ironically, has been misinterpreted as indicating a belief in rigid genetic 'determination'.

There are some human behaviour patterns for which, it seems pretty clear, not too much re-writing is needed. However complicated by civilization and culture our social life may be, not even the most implacable foe of 'sociobiology' would object to a Darwinian answer, in terms of gene propagation, to a question like 'Why do we have sexual desires?' The foe would have to be very naive, as well as implacable, if he or she objected to a Darwinian answer on the grounds that most people use contraception with the deliberate intention of preventing gene propagation. With the moth example fresh in our minds, there is no need for me to spell out why this would be naive.

Between the easy examples like sexual desire, and the difficult examples like the supposed desire for bigger desks, there is a range of intermediate examples which are legitimate subjects of controversy. Incest avoidance is one such inter-mediate case. There is a very good case for thinking that one reason why so many humans avoid mating with close relatives is that natural selection of their ancestors' genes has led them to do so. Most social anthropologists, on the other hand, object to this kind of explanation, and have erected elaborate, non-genetic explanations of their own, often ascribing a so-called 'social function' to incest avoidance, although they often do not explain why something that is good for 'society' should therefore exist. Since my colleague at the Institute of Contemporary Arts debate, Patrick Bateson (e.g. 1983), has provided one of the best and most balanced sociobiological accounts of incest avoidance, I shall not attempt to do the same.

The only point I would make is that many of the so-called objections to Darwinian theories of incest avoidance collapse into nothing when simple lessons of the same kind as the lesson of the moths and the moon are taken on board. It is commonly argued, for example, that if humans really had a built-in tendency to avoid incest, say a lack of desire for members of the opposite sex known intimately since early childhood, there would be no *need* for legal or religious prohibitions, no need for a taboo (e.g. Rose et al., 1984, p. 137). Legal and religious prohibitions are only necessary, so the argument runs, to stamp out something that people want. Therefore there cannot be any built-in incest avoidance mechanisms. The reason this is an unacceptable argument is that it contains the hidden assumption that the mechanisms postulated are infallible. It is like saying the following. Cars have locks on their doors and locks on the ignition switch. The lock on the ignition switch cannot be an anti-theft device because there is no need for an anti-theft device inside the car, given that there is always a lock on the door of the car. Instead of being an anti-theft device, the need to insert a key in the ignition switch must, therefore, have a purely symbolic, ritual significance, or a 'social function'. The fact is, of course, that whenever you are trying to prevent something bad happening, extra barriers are always a good thing, so long as no one barrier is completely infallible. Door locks sometimes fail, people sometimes forget to lock them, windows can be smashed. Safety devices are routinely redundant, backed up. Things can go wrong.

In the case of incest avoidance, it is a manifest fact that things sometimes go wrong. Incest does happen. Ironically indeed, this fact is sometimes brought forward as a debating point *against* the view that incest avoidance is favoured by natural selection! In spite of religious and legal prohibitions, and in spite of whatever built-in incest avoidance mechanisms there may be, incest still sometimes occurs. Far from two mechanisms being too many, we could, in fact, benefit from yet a third mechanism for avoiding inbreeding. It is another and very difficult question whether, and how, in the sense of proximal mechanisms, genetic selection could have led to codified anti-incest laws or religious practices. But because it is a difficult question, this is no excuse for the dismissive tone levelled at the very idea of attempting to answer it (e.g. Rose et al., 1984, p. 137). Bateson (1983) gives a thoughtful and sympathetic discussion of possible answers.

Another argument that we sometimes hear, and which I should deal with briefly, is this. Incest taboos vary so much from culture to culture – even the definitions of kinship terms vary so much – that incest avoidance cannot be due to natural selection. This is exactly as sensible as the following argument: the desire to copulate cannot have been built in by natural selection, because some people use the missionary position while others do it dog-fashion. Or as saying that the desire for food cannot have been built into us by natural selection, because different cultures define and classify food in radically different ways.

I should now take up the matters of 'reductionism' and 'biological determinism'. This is not because they are particularly interesting (they aren't; I should have preferred to spend the time discussing the genuinely interesting, if rather advanced and specialized, disagreement that Patrick Bateson has with me over units of selection), but because they occupy such a prominent place in the arguments of 'radical scientists', for example Rose et al., from whom the following representative quotations are taken.

Reductionists, according to Rose et al.,

> try to explain the properties of complex wholes – molecules, say, or societies – in terms of the units of which those molecules and societies are composed. They would argue, for example, that the properties of a protein molecule could be uniquely determined and predicted in terms of the properties of the electrons, protons, etc., of which its atoms are composed.

So far, with the usual ritual bow towards quantum indeterminacy, I am a reductionist and proud of it. But mark the sequel:

> And [reductionists] would also argue that the properties of a human society are similarly no more than the sums of the individual behaviors and tendencies of the individual humans of which that society is composed. Societies are 'aggressive' because the individuals who compose them are 'aggressive', for instance.

Since I am described elsewhere in the same book as 'the

most reductionist of sociobiologists', I must assume that Rose et al. seriously think that I would subscribe to the belief that 'Societies are "aggressive" because the individuals who compose them are "aggressive" ', or more generally to the belief that words like 'aggressive' can necessarily be applied in the same way to large units (such as societies) and their parts (such as individuals). I do not, of course, subscribe to this ridiculous belief, and I question the good faith of Rose et al. in implying that any serious scientist does. The belief attributed to 'reductionists' is exactly equivalent to the following: 'A bus drives fast, because the passengers sitting inside it are all fast runners.' Can Rose et al. really not see the difference between absurdities of this kind on the one hand, and the perfectly reasonable belief on the other hand that the properties of complex wholes can be explained in terms of the units of which those complex wholes are composed? I may believe that Patrick is a clever man, and also that all of his properties, including his cleverness, are ultimately to be explained in terms of his constituent parts, without committing myself to the belief that every atom of his constitution is a clever atom! The only author I can think of who even comes close to being a reductionist in this naive sense is the theologian Teilhard de Chardin (ironically, regarded by his disciples as an anti-reductionist champion), who thought that since conscious humans are made of atoms, there must be 'some sort of psyche in every corpuscle' (trans. 1959, p. 301).

It seem then that, for all that reductionism is one of their favourite words of abuse, Rose et al. have not thought very carefully about what they mean by it. Indeed, quite apart from the two meanings mentioned above, they seem to use it in at least two further senses. For instance (p. 160), the drawing of analogies between the social behaviour of humans and baboons is described (rather snobbishly) as 'reducing' human social biology to that of baboons. In other places reductionism seems to mean the seeking of only a single cause for a phenomenon that really has many interacting causes; in this sense it is clearly reprehensible but in this sense sociobiologists are not, as a class, guilty of it. Sometimes Rose et al. seem to be using reductionism essentially as a synonym for analysis or explanation, and it is then not clear what the objection is. As the Medawars (1984), after remarking that 'Reductive analysis is the most successful research strategem ever devised', put it:

> some resent the whole idea of elucidating any entity or state of affairs that would otherwise have continued to languish in a familiar and nonthreatening squalor of incomprehension.

The important thing about 'reductionism', it would seem, is to be against it – like Sin – whether or not you have a clear idea of what you mean by it. We had better pick one meaning of the word, in order that we can have a sensible discussion about reductionism. I shall adopt Rose et al.'s first definition, and happily call myself a reductionist in the sense of one who tries to explain the properties of a complex whole in terms of (*not* 'as the sum of': the explanations can involve highly complex interactions and require the full power of cybernetic mathematics) its constituent parts. Indeed, the only alternative to reductionism, in the sense in which I admit to being a reductionist, is religious mysticism. But we have to think carefully about how we shall actually proceed in our reductionist explanations. I shall make a distinction between two strategies of reductionist explanation, to be called 'step-by-step reductionism' and 'precipice reductionism'. Precipice reductionists probably do not exist in the world of real scientists, but they have to be mentioned because they are frequently set up as straw men. Step-by-step reductionism is the policy adopted in practice by all scientists with a sincere wish to understand what is going on.

Suppose that we wish to understand the workings of a complex entity such as an electronic computer, a human brain, or a society. The form of the argument is the same for any complex entity, and I shall choose the digital computer as my representative example. All but mystics would agree that the workings of the computer are in principle explicable in terms of the laws of physics, in this case the laws governing the behaviour of electrons in conductors and semiconductors. But only the most foolhardy of precipice reductionists would attempt a detailed semiconductor explanation of, say, a recursive procedure that analyses the syntax of an English sentence. It is not that the semiconductor explanation is actually wrong. It is just that it is hopelessly impractical when the phenomenon to be explained *emerges* from complex *interactions* of units which are themselves at a far higher level of hierarchical organization than the semiconductor devices. The step-by-step reductionist acknowledges this, and never attempts an explanation in terms of entities more than one or two steps below the hierarchical level of the phenomenon to be explained. In the case of the sentence-parsing program, the appropriate level of explanation is in the realm of software, and the units of explanation are not semiconductor devices but 'subroutines', 'lists', 'stack pointers', 'return addresses' and the like. Having made use of these units of explanation, at the level immediately below that of the phenomenon to be explained, the step-by-step reductionist is then in a position to drop down to that lower level and repeat the procedure. What were previously units of explanation – subroutines, stack pointers, and the like – now become objects of explanation in their own right, and the units of explanation are sought at the next level down. Eventually, in this orderly process of hierarchical descent, step-by-step reductionists reach semiconductor devices, and then even deeper levels of physics, but they descend the hierarchy in a gradual and a controlled way. The precipice reductionist (if one really existed) would try to go all the way in one step – jumping over the precipice rather than climbing down the staircase – and would, of course, fail.

When a computer goes wrong, a repair engineer who is a step-by-step reductionist (they always are in practice), would try to isolate the major part of the computer in which the fault lies before narrowing it down by successive steps, until reaching a unit that can be replaced economically, say a faulty chip or component-board. The precipice reductionist (who does not really exist) hangs an oscilloscope lead on every one of the billion connections in the computer, and dies of old age before finding the fault. The anti-reductionist engineer (who also does not really exist) blames the 'whole' computer and throws the whole computer away.

Reductionism, then, has become a fashionable dirty word, but attacks on it obscure an important distinction. Precipice reductionists would deserve to be attacked if they existed, but in my experience they do not. Step-by-step reductionists not only exist; their ranks include all successful scientists, even the 'radical scientists' whose favourite pastime is attacking what they call reductionism; even J. D. Watson, for all that he sounded a little like a precipice reductionist in his talk in this series. Step-by-step reductionism, including 'software' explanations' (Dawkins, 1976) and all the sophisticated techniques of cybernetics and systems theory as well as more obvious 'hardware explanations', is the only practicable, non-mystical approach to the understanding of complex entities; not just computers but cells, organisms and societies as well.

The second straw man that needs to be dealt with is the genetic 'determinist'. Above everything else, sociobiologists are accused of genetic determinism.

> Sociobiology is a reductionist, biological determinist explanation of human existence. Its adherents claim, first, that the details of present and past social arrangements are the inevitable manifestations of the specific action of genes (Rose et al., 1984, p. 236).

As far as I am aware of the views of actual sociobiologists, the latter statement is simply false. To be more charitable it may be wishful thinking: some people have a compulsive need to exorcise devils, and if the devils do not exist they must be conjured up artificially (Medawar and Medawar, 1984, p. 144). With the possible exceptions of the botanist C. D. Darlington, and one eminent American anthropologist whom I shall not name as he is alive, I do not think I have ever met a reputable scientist who would assent to that statement about inevitable genetic determination of human social arrangements. Fortunately I have discussed this whole myth of inevitability in detail elsewhere (Dawkins, 1982, Chapter 2), and Bateson deals with it in this volume, so I can be brief here.

Firstly, the majority of scientists today, including 'radical scientists' who, after all, mostly call themselves materialists, are determinists in the philosophical sense that they believe that everything that happens is caused by events in the past, and that 'free will' is a product of complex causation rather than of no causation. This does *not* mean, however, that any one class of determinants (e.g. genes) are pre-eminent in that their effects cannot be over-ridden or reversed by other determinants. Secondly, sociobiologists are more than just materialist scientists; as we have seen, they are Darwinians. Anybody wishing to offer a Darwinian explanation for some biological phenomenon has to postulate genetic variation in that phenomenon, otherwise natural selection could not have led to its evolution. This is why sociobiologists frequently postulate 'genes for' x or y. Whatever x may be, we have to postulate genes for x if we want to discuss the possibility of the Darwinian evolution of x. But the implication of a specific *inevitability* in genetic determination; of a determination that is somehow *more* 'deterministic' than the ordinary determinism that all non-mystics subscribe to; the belief that genetic determination is inevitable in the sense that it cannot be reversed by environmental agents, is nowhere to be found in the writings of sociobiologists.

As we have seen, Rose et al. imply that such strong and naive determinism is an essential element in sociobiology but, if we look carefully at their direct quotations, it turns out that these are not from sociobiologists at all, but from conservative politicians and fascist journalists! 'If the Lord had intended us to have equal rights to go to work, he wouldn't have created men and women. These are biological facts, young children do depend on the mothers.' These off-the-cuff words of a British cabinet minister seem to owe more to Christianity than to sociobiology.

> The National Front in Britain and the Nouvelle Droite in France argue that racism and anti-Semitism are natural and *cannot be eliminated*, citing as their authority E. O. Wilson of Harvard, who claims that territoriality, tribalism and xenophobia are *indeed* part of the human genetic constitution, having been built into it by millions of years of evolution (Rose et al., p. 27; my italics).

Now Wilson was discussing the interesting possibility that some tendencies towards xenophobia, territoriality etc. had evolved by Darwinian natural selection. To do this it is, of course, necessary to *postulate*, for the sake of argument, a genetic basis for the traits ('In order for a trait to evolve by natural selection it is necessary that there be genetic variation in the population for such a trait' – Lewontin, 1979). But Wilson never mentioned inevitability (the implication that the traits 'cannot be eliminated'), and he had no need to even *consider* inevitability since natural selection can perfectly well work on genetic traits that are environmentally reversible. The alleged implication of inevitability is a product of the imaginations of Rose and his colleagues, and of the members of the National Front. The word 'indeed', which I italicized in the quotation from Rose et al., is revealing: it suggests that Rose et al. are falling into the error of assuming that if xenophobia were part of the human genetic constitution, this would mean that it could not be eliminated from individual humans. Rose et al., of course, deny that it is part of the human genetic constitution, and they may well be right. But to admit that if it *were* part of the human genetic constitution it therefore could not be eliminated, is not only wrong but politically dangerous. The correct and, as it happens, politically more prudent thing to say to the National Front would be: It does not matter a damn whether xenophobia is or is not part of the human genetic constitution: even if it is, this has nothing to do with whether it can, or cannot, be eliminated by education and example.

To conclude, sociobiology is defined as a neutral field of study. In practice most of those scientists that are commonly (if sometimes unwillingly) called sociobiologists do, as a matter of fact, agree about a number of things, principally that it is worthwhile attempting to apply Darwinian ideas to behaviour. Those points of agreement do *not*, however, include any particular view about the relative 'determinacy' or irreversibility of genetic as opposed to environmental determinants of development. Nor are sociobiologists any more 'reductionist' than scientists in general: they practise the same kind of 'step-by-step reductionism' as all successful modern scientists. There are, of course, many debates and points of disagreement within sociobiology and related disciplines, and some of these are of great interest, for example the argument raised in the accompanying chapter by Patrick Bateson over genes as units of selection. But 'The Sociobiology Debate' is a trumped-up, ideologically inspired, time-wasting storm in a teacup.

7

Sociobiology and Human Politics

Patrick Bateson

Most of my concern about sociobiology is about what it seems to be in the minds of casual onlookers, rather than what the biologists interested in the evolution of social behaviour are really up to. Admittedly, some hard-selling proponents of sociobiology have been intent on using their relatively small portion of evolutionary biology as a means of taking over the human social sciences. This zeal makes them unpopular with other academics who feel threatened or insulted, but is hardly a matter of concern to the public at large. When the prosely-tizing sociobiologists are muddled (as they often are), they will eventually be sorted out in the course of discussion with their colleagues; and inasmuch as they have a point to make, it should be heard. However, some sociobiological views have been known to appeal to people with a strong interest in maintaining their power and privileges.

The differences between people are often thought to be adaptations, the product of Darwinian evolution and, therefore, attributable to genetic differences. To the non-biologist 'genetic difference' implies inevitability – which is where the trouble starts. People, who are clearly exploited or oppressed, are told that they should accept their lot because they can do nothing about their genes. Such views, though not usually shared by the scientists who seem to give these views credibility, are now a part of our political life. For that reason and perhaps unfairly, genetic determinism has loomed large as an issue in most public discussions of sociobiology.

In this chapter I shall first discuss what is meant by 'gene' and what is selected in the course of biological evolution. These are issues on which Richard Dawkins and I have different views. The debate about the selfish gene will lead me into a discussion of competition and co-operation. The issue has important political implications, because what is natural is often taken to be right and the 'struggle for existence' was thought to be natural. The result has been that, in public debate, the ideal of social co-operation is often treated as high-sounding flabbiness, while individual selfishness is regarded as the sole basis for a realistic ideology. I then turn to the matter of genetic determinism, since I agree with Rose et al. (1984) that the political implications are not trivial. Nor do I think that all sociobiologists can altogether escape the charge that they have been naive about the ways in which genes affect behaviour. The trouble with the debate about genetic determinism is that it has distorted perception of the interesting role that bio- logists' ideas can, nevertheless, play in the social sciences, so I shall pick this point up and consider it in relation to the explanation of human sexual preferences. Finally, I consider how biological ideas have been misused but can also be applied fruitfully in dealing with the arms race.

What is the Selfish Gene?

Dawkins' (1976) image of the selfish gene has made him justifiably famous. He was clearly and deliberately using a rhetorical device when he attributed motives to genes. He obviously did not think that each gene had a mind of its own. His device, which I shall consider further below, worked in the sense that he was able to illuminate aspects of evolutionary theory that otherwise seemed opaque. For all that, the precise meaning of 'gene' slithered around a lot in his writing. When I raised the issue of definition (Bateson, 1982), Richard Dawkins replied that he used 'gene' in the same way as the population geneticists (Dawkins, 1982). If so, a genetic entity (or strictly an allele) that transmits its influence from parent to offspring must be defined in relation to another one that transmits influence of another sort. The genetic difference is identified by means of a biochemical, physiological, structural or behavioural difference (after other potential sources of difference have been excluded by appropriate procedures). Unfortunately, this perfectly respectable operational definition is often replaced by a highly misleading shorthand. The gene that contributes to an individual animal behaving differently from others becomes the gene *for* its distinctive behaviour. Even the professional population geneticists sometimes forget that they are using a shorthand. For instance, Charlesworth (1978) referred to the gene *coding* for altruism. The implication is that the gene represents the behaviour and that a one-to-one correspondence will be found between them. If this were true, as it is when a gene makes a protein, the gene can be properly treated as an absolute unit and insistence on referring to differences would be mere pedantry. However, in most cases the slippage in meaning is unjustified.

Dawkins is aware that he uses 'gene' in distinctly different ways. However, he suggests that his move backwards and forwards between the orthodox language of genetic differences and the language of gene intentions is like the switching in and out of a Necker Cube (Dawkins, 1982). The lines representing the edges of a cube can be seen as though either the top corner of the cube is facing out or it is facing away. Both perceptions are equally valid. At first the Necker Cube analogy seems appealing, but it is not exact. As we have seen, in the technically correct language of population geneticists, a genetic allele must be compared with another from which it differs in its consequences. In selfish-gene language, it stands alone as an entity, absolute in its own right. The perception generated by one meaning of gene does not relate to the same evidence as that generated by the other. Both Richard and I may be selfish, but the difference between us certainly is not. You cannot attribute motives to a comparison.

Consider the problem when, in the course of evolution, an allele, properly defined in terms of a difference, goes to fixation and all members of the population have a double complement of the allele. The other alleles, with which it was being compared, have ceased to exist and so has the relative basis for defining this gene. Of course, the DNA which made the crucial difference survives, but quite different scientific operations are required to demonstrate its existence. This makes me think that Richard Dawkins uses a double standard. He falls back on a relative definition of genes when pressed to be precise, but retains an absolute definition for his selfish gene talk. This would not matter if it did not cause so much muddle and encourage the genetic determinism, which both of us find so misleading. Richard Dawkins's notion of an absolute genetic unit, supported by the unfortunate genes-for-characters shorthand, made him *look* like a genetic determinist. Moreover, it led him to make what I consider to be a mistake when he redescribed the Darwinian theory of evolution.

Levels of selection

Charles Darwin's (1859) theory required that three conditions must have held if adaptation to the environment occurred in the course of biological evolution. First, variation must have existed. Second, some variants must have survived more readily than others. Third, the variation must have been inherited. Darwin used the metaphor of selection to describe the evolutionary process of adaptation because he had in mind the activities of human plant- and animal-breeders. A person who wants to produce a strain of pigeons with longer tail feathers than usual, picks from the flock those birds that have the longest feathers and exclusively uses them for breeding purposes. This is artificial selection. By analogy, Darwin referred to the differential survival of the characters that adapt an organism to its environment as *natural* selection.

Richard Dawkins has frequently pointed out that individual organisms do not survive from one generation to the next, while on the whole their genes do. So it seemed logical to propose that natural selection acts on what survives, namely the genes. His approach to evolution was presented in characteristically entertaining form when he suggested that the organism is '. . . a robot vehicle blindly programmed to preserve its selfish genes' (Dawkins, 1976). In order to understand what he was doing here, it may be helpful to forget biology for a moment and think about the spread of a new brand of biscuit in supermarkets. Consider it from the perspective of the recipe. While shoppers select biscuits and consume them, it is the recipe for making desirable biscuits that survives and spreads in the long run. The word in the recipe that makes the difference between a good and a bad biscuit is, in Richard Dawkins' sense, selfish, because it serves to perpetuate itself. This novel way of looking at things is unlikely to mislead anyone into thinking that what shoppers really do in supermarkets is to select a word in the recipe. By contrast the selfish gene approach has managed to run together the crucial differences between individuals which are most likely to survive with the genetic consequences of differential survival in later generations. The genes have been, in my view, erroneously promoted as the units of natural selection.

I doubt if Darwin would have confused the crucial surviving character with the means by which it recurs again in the next generation. Nevertheless, confusion has evidently arisen and, therefore, an explicit distinction should now be made between the cause of differential survival and its effects on the frequency of the genetic replicators. Once made, the distinction saves a good deal of muddle in modern discussions of evolution. It also serves a valuable role in drawing attention back from a preoccupation with single genes to the ways in which genes work together. As Richard Dawkins would be the first to admit, the long-term survival of each gene depends on the outcome characteristics of the whole gene 'team'. Furthermore, special combinations of genes will work particularly well together, and the gene that fits into one combination may not fit into another. Bringing back the old notion of the *coadaptation* of genes is helpful in re-establishing that organisms do, indeed, exist as entities in their own right.

It seems likely that when particular combinations of genes making up a team became important, characteristics of the organism that helped to hold the combination together almost certainly evolved. This process would have been given an interesting and complicated twist in sexually reproducing organisms, since the method of reproduction creates variation by recombining genes in each generation. Some variation is presumed to be beneficial, otherwise sexual reproduction would have disappeared. However, certain features of behaviour set a limit on the extent of genetic recombination (and consequent variation among offspring). Particularly striking are the behavioural preferences of animals which may actively avoid mating with individuals very different from themselves. Japanese quail do not mate with birds of a plumage type markedly different from the birds with which they have grown up. When reared with siblings and then offered a variety of differently related members of the opposite sex to choose between as adults, quail prefer cousins (Bateson, 1983a). So their behaviour looks as though it evolved to minimize the costs of extreme inbreeding while, at the same time, holding together in the offspring combinations of genes that work well together.

Another useful line of thought flows from separating differential survival by natural selection from genetic transmission. Once we have distinguished the well-adapted character that survives from one generation to the next from the means by which it recurs again, we can ask: to what does that character belong? Does it have to be the property of an individual? I suspect that in attempting to answer the question, Richard Dawkins is ambivalent. On the one hand he wants to attack the confused group selectionist argument of characters being good for the species. On the other, he has cleverly shown how the phenotype of an individual's genotype extends outwards into the physical environment and the bodies of other organisms. What is missing, though, is a clear acceptance that the character, on which natural selection acts, may be formed by more than one individual (Alexander and Borgia, 1978). The lichens provide marvellous examples of symbiotic co-operation.

Lichens are found everywhere from the Arctic to the tropics – and on virtually every surface from rocks and old roofs to tree trunks. A spectacular example is Spanish moss which hangs down from the branches of trees in warm places. They look like single organisms. However, they represent the fusing of algae and fungi working together in symbiotic partnership. The characteristics of the whole entity provide the adaptations to the environment. One lichen, acting as an organized system, could compete with another in the strict Darwinian sense of differential survival. The crucial point here is that the character enabling one lichen to survive better than the other is an emergent property of the two species living in symbiotic relationship.

A similar point can be made about the mutualisms and reciprocated aid of co-operating animals. Many birds and mammals huddle to conserve warmth or reduce the surface exposed to biting insects. Male lions co-operate to defend females from other males. Mutual assistance is frequently offered in hunting; for instance, co-operating members of a wolf pack will split into those that drive reindeer and those that lie in ambush. As a result they all get more to eat. In highly complex animals aid may be reciprocated on a subsequent occasion. So, if one male baboon helps another to fend off competition for a female today, the favour will be returned at a later date. What is obvious about such cases is that all the participating individuals benefit by working together.

A more subtle issue is that the outcome of their joint action could have become a character in its own right, like an overall well-adapted feature of a lichen. The character generated by the mutualistic arrangements could distinguish one social group from another and become something on which natural selection would have acted in certain circumstances. These would arise when cheating individuals were penalized. Clearly a cheat could sometimes obtain the benefits of the others' co-operation without joining in itself. However, such actions would not be evolutionarily stable if the cheat's social group was less likely to survive than a group without a cheat *and* the cheat could not survive if it left its own social group.

Just as with coadapted complexes of genes, once evolutionary stability was achieved, it is likely that features evolved to maintain and enhance the coherence of the highly functional co-operative behaviour. Signals that predict what one individual is about to do, and mechanisms for responding appropriately to them, would have become mutually beneficial. Furthermore, the maintenance of social systems that promoted quick reading of familiar individuals would have become important. If this happened, natural selection would have acted on the outcome of the joint actions of individuals in the social group. To adopt such a view is not to return to muddled good-for-the-species thinking. It simply requires acceptance that the characteristics of social groups are the emergent properties of the participating members and the logic of Darwinian theory applies as much to these characters as it does to those of individual organisms.

The general conclusion is that what constitutes an adaptive character on which natural selection can act is not be found at any one level of organization. The character could be at the level of DNA if one molecular configuration can successfully displace another. It could be at the familiar level of individuals' characters or, as I have suggested here, at the higher-order level of social structure. These important points about biology and, I might add, human society are lost if evolution is exclusively thought of in terms of what is good for the gene. This brings me to a political matter.

Competition, co-operation and politics

The emphasis on selfishness and the struggle for existence in evolutionary biology has had an insidious confirmatory effect on the public mind (Bateson, 1984). Competition has been widely seen as the mainspring of human activity, at least in Western countries. Excellence in the universities and in the arts is driven by the same ruthless process that supposedly works so well on the sports field or market place. Individuals thrive by competing and winning. This view of human nature, currently so popular among right-wing politicians, has been justified by an appeal to biology and unwittingly the biologists were, in their turn, somewhat influenced by the public mood. The image of selfish genes fused imperceptibly with the ideology of individualism.

None of us notices everything, and our tendency to draw foolish generalizations is subject to quick correction by others whose experience has been different. The corrective process slows down, though, when everyone has been induced to look in the same direction at the same time. This can happen when people are jointly persuaded by a powerful set of ideas. It has often been noted that Darwin himself was strongly influenced by the economic theories of the free market that stemmed from Adam Smith. Furthermore, the return of Darwin's ideas from biology to the wider domains of social and political thought was almost certainly fostered by the commercialism of nineteenth-century Britain. This does not deny the scientific value of the ideas, but explains why certain facets of them were so rapidly and widely assimilated. The emphasis was heavily on the cleansing role of social conflict. As Kropotkin (1902) complained at the time, neither the biologists nor those who found their ideas congenial focused as much as they might have done on examples of mutual aid. Eighty years later, Maynard Smith (1982) observed that the importance of two animals co-operating, because it paid both of them to do so, was under-rated in the initial writings of sociobiology. The position has changed radically in the past few years and the professional literature is now rapidly accumulating articles on symbiosis, reciprocity and mutualisms (Wrangham, 1982). The message has not yet percolated extensively into public discussions of human social behaviour. It is very important that it should.

As things stand, the appeal to biology by the New Right is not to the coherent body of scientific thought that does exist but to a confused myth. Biology is thought to be all about competition – and that means struggle. The Darwinian concept of differential survival has nourished the belief in the importance of individualism. In fact, there has been a fourfold muddle.

1. Competition in the past does not necessarily mean competition now. As in the ideal and unfettered free market, unrestrained competition within an ecological niche can generate a monopoly.
2. Differential survival is not equivalent to social conflict. One individual may be more likely to survive because it is better suited to making its way about its environment and not because it is fiercer than others.
3. Selfishness is not incompatible with co-operation. Individuals may survive better when they join forces with others. By their joint actions they can frequently do things that one individual cannot do. Consequently, those that team up are more likley to survive than those that do not.
4. Individuals cannot be abstracted from society. Moreover, their actions may contribute to the survival of the social group of which they are a part.

I do not wish to imply for one moment that social conflict never occurs. Clearly co-operative arrangements do break down and frequently individuals do compete for the same food, the same mate, the same place to nest or a host of other valued resources. However, it is simply not true that animals never depend on others or never reveal what they are about to do next. They often do, and it is often in the interests of both parties that communication between them should work effectively. Furthermore, the importance of behaving honestly in co-operative arrangements is crucial. Cheats wreck their own chances of survival if they destroy the basis on which members of a social group can work together with trust.

The political danger of representing all human social relationships in terms of competition is that the expectation is self-fulfilling. Trust, which is a necessary condition for willing co-operation, is poisoned. Without trust, how do you get people to perform the concerted activities that are required in a modern industrialized state? You buy, manipulate or coerce them – all deeply alienating in the end. Within a democratic state, a crude competition model of social behaviour destroys the basis by which people work together with some confidence and pleasure. Rampant individualism is morally corrupting and socially divisive.

Genetic determinism

The great majority of biologists would agree that the evolution of life has generally involved changes in the genetic composition of the evolving organisms. A second point of consensus is that, if the change over time involves an adaptation to the environment and the postulated process of Darwinian selection has worked at all, then the genes must *influence* the characteristics of the organism. However, the disagreements start over the particular ways in which genes affect the outcome of an individual's development. Some sociobiologists (though not Richard Dawkins) seemed to think that a straightforward correspondence could be found between the genes and behaviour. E. O. Wilson (1976) was quite candid about it when replying to reviewers' criticisms that he was naive about behavioural development in his book, *Sociobiology: The New Synthesis* (Wilson, 1975):

> Population biology, particularly the corpus of evolutionary ecology which now dominates this field, operates on the simple premise that most phenotypic traits of parents and offspring are correlated to some extent by virtue of a higher than average possession of the same genes. . . . As a rule theoretical socio-biologists and behavioural ecologists treat the complex links between gene and behavioural phenotype – for example, the exact structure of the nervous system, the full potential developmental pathways, and endocrine–behavioural inter-actions – as modules that can be temporarily decoupled from the explanatory scheme. . . . The procedure is an unavoidable consequence of the current state of the art in population biology. Let me be the first to admit that if the explanations really cannot be decoupled, and if the algorithms of develop-mental biology can be defined that somehow lie outside the framework of natural selection, then most of theoretical socio-biology will collapse, and with it basic evolutionary theory.

The final admission was handsome but unnecessary. The explanations for the 'decoupled' developmental biology are irrelevant to, and at a quite different level from, evolutionary explanations – a point I shall return to shortly. However, what the quotation reveals is Wilson's belief that the development of the individual is merely a complex process by which genes are decoded. If genes code for structures or behaviour patterns they must bear a straightforward relationship to them.

An alternative view of an individual's development is clarified by a culinary metaphor which both Richard Dawkins and I have used. When I first wrote about it (Bateson, 1976), I was attacking the view, found in classical ethology (Eibl-Eibesfeldt, 1970; Lorenz, 1965), that learned and unlearned components are intercalated in adult behaviour and that the unlearned bits are represented in the 'genetic blueprint'.

> Is it really possible to break up the fully developed song of an experienced male chaffinch into components, some of which are specifically affected by experience and some of which are not? Even though we know that many factors have been responsible for the detailed specification of the song (Thorpe,

1961), it does not follow that somehow these factors will correspond to the constituents of the final behavioural product. Rather than liken the development of such behaviour to the insertion of days into an existing calendar (*intercalare*), I suggest a more appropriate analogy would be the baking of cake. The flour, the eggs, the butter, and all the rest react together to form a product that is different from the sum of the parts. The actions of adding ingredients, preparing the mixture, and baking all contribute to the final effect. The point is that it would be nonsensical to expect anyone to recognise each of the ingredients and each of the actions involved in cooking as separate components in the finished cake. For similar reasons, I think those cases in which a simple relationship can be found between the determinants of behavior and the behavior itself will be exceptional.

Five years later, Richard Dawkins (1981) likened the cake's recipe to the genetic code. He put it as follows:

> The genetic code is not a blueprint for assembling a body from a set of bits; it is more like a recipe for baking one from a set of ingredients. If we follow a particular recipe, word for word, in a cookery book, what finally emerges from the oven is a cake. We cannot break the cake into its component crumbs and say: this crumb corresponds to the first word of the recipe; this crumb corresponds to the second word in the recipe, etc. With minor exceptions such as the cherry on top, there is no one-to-one mapping from words of recipe to 'bits' of cake.

In many ways the implication that all the instructions for making organisms are to be found in the genes is unfortunate (Partridge, 1983). Indeed, Susan Oyama (1985) has argued eloquently against the use of metaphors that attribute so much to genes. However, the point on which she, Richard Dawkins and I all agree is that genes rarely if ever code for the characteristics of the whole organism.

The modern view about development is that the processes involve multiply influenced systems with properties that are not easily anticipated, even when all the influences are known. Like many artifical systems, developmental processes are also strikingly conditional in character, particularly those in complex organisms. In one set of conditions they proceed in a particular and appropriate direction, in another set they do something different but equally appropriate. This means that the expression of genes and the characteristics they influence are not inevitable. The overall performance of the individual is usually exquisitely tuned to prevailing circumstances. For instance, some usually green grasshopper species growing up on African savannah blackened by fire are black (Rowell, 1971). As a result they are less easily detected by predators than if they had been green like their parents. Their descendants, developing among new grass, suppress the mechanisms making black cuticle, and once again are green. They too are not easily detected by predators and once again the appropriate decision has been taken during development. If, as seems likely, expression of particular genes is required (though not sufficient) in order to make the cuticle a particular colour, some genes will never be expressed in the lifetime of an individual.

The evolution of ability to track the environment did not stop with simple conditional responses of the type seen in the African grasshopper. With the elaboration of nervous systems, came more and more powerful rules for tracking the environment by learning. The value for the animals is obvious. The implications for the way we think about such an animal are also crucial. The dependence of the animal's behaviour on external conditions means that it will be no more possible to predict precisely what an animal will do from the knowledge of its genes than it will be possible to predict the detailed course of a game of chess from the knowledge of the game's rules and what the pieces look like.

Behind E. O. Wilson's uncertainty about the nature of developmental processes lay a straightforward muddle between evolutionary and developmental arguments. At the heart of much of his writing lay a category mistake – a running together of two quite different issues. Wilson and his supporters were primarily interested in evolutionary arguments and the biological functions of particular forms of social behaviour. These arguments and styles of thought have a role to play in the understanding of human social behaviour – a matter that I shall discuss in the next section. However, just how the behaviour develops in individuals should have been left as an open question. Whether or not development involves some 'instruction' from a normally stable feature of the environment, or whether it would be changed by altering the prevailing social and physical environment, cannot be deduced from even the most plausible evolutionary or functional argument. Wilson's synthesis of what was known (and guessed) about the evolution of social behaviour had no bearing on whether or not human behaviour can be changed by altering the social or physical environment.

The charge of naiveté about what happens as an individual develops was as damaging to sociobiology as it had been to classical ethology. It was particularly important to Wilson that he should respond to the charge when writing about the manifest plasticity of human behaviour. His solution was to replace genes that prescribe the form of behaviour by genes that do the same for 'epigenetic rules'. These hypothetical rules are supposed to determine how development proceeds and how learning takes place. However, E. O. Wilson's notion of how the rules themselves develop was markedly different from the thinking of the major theorists in developmental biology such as Waddington. With Lumsden he defined epigenetic rules as 'genetically determined procedures that direct the assembly of the mind' (Lumsden and Wilson, 1981). Clearly an important aspect of development was still decoupled – at least in some assembled sociobiological minds. Furthermore, the charge of genetic determinism still stuck.

I should point out that many of the thinkers who have been lumped with E. O. Wilson have thought much more clearly than he did about the issues. Long before *The Selfish Gene*, Richard Dawkins (1968) had written one of the most sophisticated discussions of the time about the interplay between genes and the environment during development. His most recent book (1982), contains the best rebuttal of genetic determinism that I have come across. Furthermore, the majority of the scientists who call themselves sociobiologists are now aware of the necessity of separating functional and evolutionary arguments from developmental ones.

Biological evolution and social prescription

In the polarized state of opinion stemming from the mistaken emphasis on genetic determinism, it was not easy to renew the case for introducing some evolutionary thinking into the human social sciences. Nevertheless, it was worth another try because the promise of fruitfully combining the insights of different disciplines remained real. In an outstanding essay, Hinde (1984) showed how an approach to sex differences in humans in terms of biological function binds together and makes sense of what is observed without implying that the differences are inevitable, unchangeable or even desirable in a modern context.

An equally sensitive issue is whether or not evolutionary thinking can add anything to our understanding of social prescriptions about mates. In general the biologists' contributions on this issue have not been of the same calibre as Hinde's. Indeed, many commentators have thought it was enough to point out that incest taboos exist because they generally have the effect of preventing inbreeding (van den Berghe, 1983). Since inbreeding has biological costs and incest taboos reduce the chances of incurring them, it was supposed that taboos must be a product of natural selection and therefore function to increase reproductive success. However, a lot remained poorly explained. In particular, how did behaviour come to be culturally transmitted as a prohibition from one generation to the next?

The position was further confused by the introduction of a different set of observations, namely that people of the opposite sex who have grown up together are not usually sexually attracted to each other as adults, even when they are not related and are encouraged to marry. The inhibited sexual preferences would usually operate against close kin, since they would normally be most familiar. As a consequence this mechanism would also serve to prevent inbreeding. Note that two different mechanisms for reducing the costs arising from inbreeding have been proposed, one involving an explicit prohibition and one an unconscious inhibition. Redundant mechanisms come as no surprise to a biologist; birds can navigate by means of magnetic cues, the sun, and the stars as well as by using landmarks such as mountains and coastlines – and even then, they sometimes get lost. Nonetheless, it is not good enough to argue that merely because inbreeding avoidance and incest taboos have the same effect, they must therefore have arisen in the same way by natural selection.

The explanation for incest taboos in terms of biological function does not to appeal to many anthropologists. It has too many loose ends and too much is left unexplained. For instance, this particular argument tells us little about the variation between cultures or the ways in which each culture arose. Alexander (1980) has made a concerted effort to meet the objections of social scientists and I have suggested elsewhere (Bateson, 1983b) that, on the specific issue of linking inbreeding avoidance with incest taboos, the possibilities for a useful dialogue between disciplines may be greatly improved if insights from biology, psychology and anthropology are used to bind the evidence together. The historical argument can be summarized as follows.

1. Humans evolved like many other animals so that experience obtained in early life provided standards for choosing mates when they were adult. Normally the experience was with close kin so that, when freely choosing a mate, a person preferred somebody who was a bit different but not too different from close relations. This system evolved because those who did it avoided the maladaptive costs of both extreme inbreeding and too much outbreeding. Consequently, the system was more likely to be represented in subsequent generations. This step could have occurred long before language evolved.

2. For other reasons to do with the benefits of cohesion, conformity was enforced on all members of the social group. Those who did things that the majority would not do themselves were actively discouraged. In sexual affairs, pressure was put on people who interested themselves either in individuals who were very closely related or who were strange to members of the group. As language evolved, prescriptions about mating were transmitted verbally to the next generation. In this way taboos and marriage rules characteristic of a culture came into existence.

This hypothesis, which was first outlined by Westermarck (1891), is deliberately stripped down to make two deductions plain. First, while prohibitions are broadly correlated with inhibitions, they did *not* arise directly as the result of natural selection in biological evolution. Second, cultural differences in prohibitions should be related to the categories of persons who are familiar from early life. So if children grow up with some types of cousin but not others, the most familiar cousins should then be prohibited as sexual partners, whereas the others should not. The same goes for non-blood relations. This prediction is testable by comparing different cultures. The hypothesis does not preclude the occurrence of incest, which certainly seems to be much more frequent in modern society than had been realized or admitted. However, it does predict that stable incestuous relationships will be most common between partners who had relatively little contact with each other during early life.

The biological-cum-psychological hypothesis is silent about the role of power, property and countless other factors in the formulation of culturally transmitted marriage rules, but obviously it does not (and should not) exclude them. Although I am critical of the excessive claims sometimes made for human sociobiology, it will be plain that I do see value in studying human social behaviour at many different levels. When ideas about what behaviour is for are brought to bear on humans, they expose a multi-layered problem.

All of us can do something that is highly functional in the sense that it helps us to achieve a goal which we have specified for ourselves. Alternatively, our actions can work in the service of reaching a goal of which we are not even aware. Clearly the goal could be specified by another person, but it could also be specified by different historical processes of selection. Culturally transmitted behaviour that worked in the past is more likely to exist in the present than behaviour that was less effective. A similar point can be made about the performance of behaviour arising from natural selection during the course of evolution. Historical processes of selection can leave their impact in the present without involving an awareness of goal. A baby that copies its mother's actions does not need to know why she performed them. Nor, come to that, need she. Similarly a great array of human actions – breathing, sleeping, making love – are commonly performed without much introspection. The functions of these actions are rarely at the front of our minds, and many people probably never think much about why they do what they do. The point is, then, that sometimes humans specify goals for themselves, consciously assessing the costs and benefits of their actions, but sometimes they do not. What is needed in approaching these problems, then, is constructive collaboration between biologists and social scientists and a proper respect for the insights that the different disciplines can provide. The same multi-layered approach can be taken back to the political arena. I shall now consider how that might be done.

The arms race

In the current debate about the arms race between the Western powers and the Soviet Union, it is often argued that states can only achieve stability through strength. Associated with this belief is a view that peace in Europe has been maintained since 1945 by the existence of the nuclear deterrent. This strikes me as an astonishingly complacent treatment of the evidence. Forty years is not a particularly long time between major wars involving European nations in the past 200 years. The gap between the Napoleonic Wars and the Crimean War was about the same. Furthermore, the time since the Second World War has been one of unprecedented prosperity both in the West and in the Soviet Union. Such relatively stable conditions are not likely to persist for ever.

Reliance on deterrence with nuclear weapons depends on human rationality. At times of international tension we know only too well how rapidly reason evaporates in the face of fear or indignation. Angry or frightened people do not make appropriate calculations about the costs of their actions. I am not, of course, suggesting that rational human beings are never deterred from aggressive acts by the thought that they might get hurt. However, I do suggest that there is a good deal more to human behaviour than rational calculations about likely benefits and costs. In special circumstances rulers and governed alike are liable to do very silly things.

What has all this to do with biology? In part, the thinking behind the arms race is based on the impoverished competition model of human behaviour which, as we have seen, relies on an inappropriate appeal to evolutionary biology. However, the view that nuclear deterrence will continue to work is also based paradoxically on an assumption that we can transcend our history. To be critical of that assumption is not, of course, to imply that some aggressive impulse will inevitably work its way out. The point is more subtle. In as much as the human readiness to fight was adapted in the past to meeting particular challenges from other human beings, the long-term benefits of behaving in this way must have outweighed the costs. The worry is that the unconscious response to provocation embodies a cost–benefit calculation that is totally out of line in present circumstances – with potentially disastrous consequences for everybody.

The problem we have to face up to is this. Whatever the historical processes that favoured war-like behaviour in humans may have been, they did not look into the future. Those processes operated in conditions that no longer apply. Furthermore, we are not necessarily aware of the ways in which our own behaviour is influenced by the action of past events. That is why people are perfectly capable of doing things that make no sense whatsoever in a nuclear age. It is questionable whether intense fear of enemies or, at the times when it is expressed, lust for war should be treated as forms of insanity. The present-day escalating arms race is generated by normal people. There is nothing obviously wrong with the mental health of our political leaders. However, their behaviour is maladaptive in the sense that it is likely to lead to their own extinction (and ours). What they have done, and are capable of doing, simply is not relevant to the changed circumstances of the present. That is probably what Einstein had in mind when he made his famous remark: 'The power set free from the atom has changed everything, except our ways of thinking.' Can anything be done about it?

Humans have a well-known ability to regard a feared stranger as inhuman and therefore as a suitable object for slaughter (Humphrey and Lifton, 1984). The other side of such fierceness is human readiness to co-operate in remarkable ways – particularly with those we know well. The irony is that our willingness to risk our own lives in destroying our enemies is part of our extraordinary ability to work for (and, indeed, die for) those whom we regard as our own. We should do well to look carefully at the conditions in which this sense of allegiance is formed and the circumstances in which the co-operation collapses.

Social life can certainly involve intense competition. It can also involve real benefits and active co-operation. The balance between these conflicting pressures can change so that if conditions become really difficult, the co-operative arrangements break down; or if members of a group do not know each other, no mutual aid may occur until they have been together for some time. As in so many other aspects of behaviour, we must operate by means of a set of conditional rules, so that the balance between competition and co-operation alters according to circumstances. But we are capable of creating those circumstances. The belief that nobody is to be trusted has an alarmingly self-confirming character to it. The conditions for working together rapidly spiral out of reach. If competition is seen as being the only mode of human existence, we have created the conditions in which that becomes true.

What we need to do as much as anything else is to work actively against a style of thought that places all the emphasis on confrontation. This corrupting and impoverished view of human nature inevitably leads to an arms race, and then inexorably to the use of those arms. Controlling the likely sources of conflict will involve an extraordinary degree of mutual understanding. However, it seems clear that the necessary co-operation will never be achieved if the present climate of mistrust persists. The conclusion is inescapable. If we wish to survive, we had better get together.

Conclusion

Applying the biologists' knowledge of social behaviour to humans has obvious political implications. However, I have come to feel that the impact of sociobiology is by no means all on the negative side. Indeed, some of the apparent support for social injustice was based on a muddle about how genes actually work. As this is straightened out and genetic determinism falls away as a serious issue, I believe that the biological knowledge can help the understanding of social issues by showing precisely how human potential is expressed in some conditions and not seen in others.

It is not necessary to appeal to biology at all, of course, and many would continue to argue that to do so remains utterly misleading and dangerous. They may well be right so long as people continue to suppose that 'natural' means 'desirable'. The line I have taken here, though, is that 'natural' by no means always means 'nasty and selfish' and, in as much as biological arguments are brought into political debates, it is appropriate that the biological value of co-operation should be fully appreciated. Whatever route is taken to political under-standing, my own hope is that we shall once again obtain pleasure and security from the trust we place in others.

Acknowledgements

I am very grateful to the following for their comments on a draft of this article: Monique Borgerhoff Mulder, Tim Caro, Robert Hinde, Deborah Hodgkin, Susan Oyama, Steven Rose, Dan Rubenstein, Peter Smith, Joan Stevenson-Hinde and Bernard Williams.

Part III
Science, Brains and Machines

8

Does Artificial Intelligence Need Artificial Brains?

Margaret A. Boden

Do you need a brain to be brainy? If by a 'brain' we merely mean some physical organ that enables its possessor to do things intelligently, then of course you need a brain to be brainy. Even if 'you' are a computer, the same applies. In this sense, artificial intelligence cannot possibly do without artifical brains.

But what if by a 'brain' we mean an organ of intelligence *that resembles the human brain*: what then? Perhaps some non-human intelligences could be brainy without one? If so, we might be able to build artificial intelligences without having to give them brains.

Let us try to get clear, first, what sort of 'resemblances' we have in mind here. We are not talking about whether or not something is made of protoplasm, nor about whether it has certain chemicals present in it. I assume that there is nothing magical about neurons: they are part of the natural world. That is, they can in principle be described by physics and chemistry, just as bone and steel – and silicon chips – can too.

I assume, also, that there is no special physical property that is essential to intelligence, which is possessed only by brain proteins. This second assumption might conceivably be mistaken – but neuroscience gives us no reason whatever to suspect that it is. Quite the contrary: the more brain scientists discover about this remarkable organ, the more they see it as a subtly complex physicochemical system.

What this all comes to, then, is that the properties of the human brain that are relevant to its function as an organ of intelligence almost certainly have nothing to do with what it is made of. Rather, they concern how it is organized and what it does.

But this is precisely what leads many people to scepticism about artificial intelligence. For most work in artificial intelligence has been done, and still is being done, on machines whose fundamental organization differs from the brain in three important ways.

First, most computers are digital systems, in which the basic units either 'fire' or they do not. The brain, on the other hand, is to a large extent an analogue device: synaptic activity varies continuously (and nerve cells often fire 'spontaneously' as a result). Second, digital computers are designed as serial devices, in which only one instruction is executed at a time. By contrast, the brain is a parallel-processing device: neurons have rich interconnections, which enable cells to encourage or inhibit their neighbours' activity. Third, digital computers are general-purpose machines, which can be used for indefinitely many problems. The reason is that they work by manipulating intrinsically meaningless formal symbols, whose meaning is ascribed by the human programmer and/or user. But many brain cells are dedicated to one purpose. The visual cortex, for instance, contains cells that respond only to lines slanting in a particular direction, and the auditory cortex too contains highly specialized neurons.

Our original question can now be restated: Can only an analogue, parallel-processing, dedicated device be intelligent? Does any system that can do the same things as the human mind can do have to have an organ which is like a human brain in these ways? Or can artificial intelligence workers ignore human psychology and physiology? Can they simply get on with the job of building some artificial thinking-organ – which might be very different from the brain?

These questions are highly controversial. In the history of artifical intelligence (if something still so young can be said to have a history), the pendulum has swung from one answer to the other – and back again. In the early days much research attempted to mimic the parallel functioning of the human brain. Then this research fell out of favour, and a quite different approach was preferred. Recently, however – and surprisingly – some of the special features of the human brain are being taken seriously again.

You may feel that this swing-back of the pendulum is not 'surprising' at all: *we* need brains to be brainy, so it is reasonable to expect that computers do too. However, this is to ignore the reasons for the recent return to brain-like computer models.

For what do we mean by 'brainy' – or by 'intelligent'? Those whom we call 'brainy' excel at abstract thought, such as mathematics, science, medicine, or even tax law. And people regarded as 'intelligent' are usually good at practical or verbal problem-solving, whether or not they are also 'brainy' in the more academic sense. Occasionally, a person's brainpower is admired primarily because of an exceptional factual memory, like that needed to compete in 'Mastermind'. In all these cases, people are marked out as intelligent because they can do things which most of us cannot.

By the same token it is simply irrelevant to our everyday judgements of intelligence that someone can see things in the world around them, or speak their native language. These mental capacities receive no admiration, since almost all of us can see and speak so well – and so unthinkingly. Even common sense is not included in the meaning of 'brainy' (so we can say without contradiction: 'lots of brains, but no common sense'). Vision, language, and common sense are remarked upon only in their absence. When present, they are taken for granted; and, being thus taken for granted, they are commonly assumed to be relatively simple.

What is surprising about the history of artificial intelligence is that these everyday capacities have turned out to be much more difficult to automate than many of the achievements of our intellectual superiors. Traditional computers – which are unlike the brain – can do pretty well those things which we do badly; but they do very badly what we all do very well.

Give a (suitably programmed) general-purpose digital computer a problem in logic, maths, chemistry, or even medical diagnosis, and it may be able to solve it. Sometimes it will do better than all but the best world experts; occasionally it will surpass even them. But ask it to see a face across a crowded room (as the song has it), and it will fail abysmally. Ordinary speech and common-sense reasoning, likewise, will be beyond it.

Some would say that this is merely because we have not yet discovered how to write the 'suitable' programs required. Certainly, if any computer could do these things, then – in principle – some serial-processing machine could do them too. However, what is true in principle may be of little practical interest – not least because of the vast length of time that would be needed to do what is possible in principle. In practice, then, a different approach may be required to automate everyday human capacities.

We already know that evolution has come up with a different approach. The crucial organs of those living machines that can talk, think, and see – namely, human brains – are organized in a fundamentally different way. This is why some workers in artificial intelligence, namely those who wish to design computers to do those things which we all do so well, are increasingly looking again to the brain.

In the process they expect to cast light on just how we ourselves do these things. For the computational processes in brain-like computers may be significantly similar to the processes going on in human brains. If so, the intellectual traffic will not be all one way: neuroscience and psychology could benefit from artificial intelligence, as well as vice-versa.

Consider computer vision, for instance. How could one enable a computer to see?

To avoid difficulties about consciousness (which would only obscure the main point being discussed), let us agree that by 'seeing', here, we will mean the ability reliably to describe things in the external world, given information provided by a visual array or image; and, if the ICA will permit such a heresy, let us ignore aesthetic appreciation: never mind beauty, what about truth? That is, we will take seeing to be the ability to use input light so as to produce accurate descriptions of things in the external world.

This descriptive ability would be an essential criterion of vision even if conscious experience were to be included also. The descriptions may be expressed in English or French, but they need not be: rats and squirrels show by their appropriate physical behaviour that in some interesting sense they can see. Further, the phenomenon of 'blindsight' suggests that my exclusion of consciousness from our discussion is not wholly perverse. Someone with blindsight lacks experience of one half of their visual field, and cannot describe verbally an object lying in it: but they may nonetheless be able to put their hand in precisely the right position (and with fingers appropriately curved) to pick up. (Perhaps some animals have visual systems enabling them to do this sort of thing, and no more.)

Our question, then, is this: How could a computer reliably report 'That's a teddy bear', or 'I don't know just *what* that thing is – but it's about a foot long, with an undulating spotted surface slanting away from the ground. And it's a couple of yards away from me (a few inches in front of the large round object)'?

The earliest work in computer vision reflected the ideas about the brain that were current at the time. People tried to build parallel-processing devices, analogous to neural networks, that would be capable of recognizing distinct visual patterns. (The fact that these 'parallel-processors' were actually simulated on serial machines is irrelevant: as we shall see, their prime weakness lay in what they did, not in the machinery with which they did it.)

One such program was called *Pandemonium*, because its authors described its action in terms of the simultaneous shouting of several information-processing units, or 'demons'. The demons were observant, but very narrow-minded. Each demon knew about, and watched out for, only one thing: perhaps a horizontal bar in the middle of the visual field, or a convex curve in the top right-hand corner. When a demon saw what it was looking for, it sent a message to the central master-demon. Each demon varied the loudness of its voice according to its judgement about the probability and the importance of its message. But each made its decisions independently: no demon could influence the loudness of its neighbours' shouting.

On the basis of the messages coming in from the various demons, the master-demon would decide what overall pattern was present. For example, the letter 'F' would be reported by the master if it was told that there was a 'mid-vertical', an 'upper-horizontal', and a 'mid-horizontal' shout. Suppose that, in fact, the vertical stroke was not *precisely* vertical: then the relevant low-level demon would whisper, rather than shout. But since no letter consists of only two horizontal bars, and since the precise tilt of the upright bar is irrelevant, the master would decide on an 'F' in this situation too.

Later versions of *Pandemonium* allowed for several levels of demons. The lowest-level demons might look for tiny segments of black in the image (for instance, a tiny horizontal line). The next level would consist of demons seeking out entire letter-strokes (several contiguous horizontal snippets could be accepted as a horizontal stroke). Next would be a level of 'letter-demons', each of which would look for a specific letter (the E-demon would demand one more horizontal stroke than the F-demon would). There might even be 'word-demons' looking for specific words (such as 'OAT', 'FAT', and 'EAT' – of which each pair differs by one letter, and the last pair are distinguished by a single letter-stroke).

There is a clear analogy between these single-minded demons and the remarkable 'feature-detector' cells in the visual cortex – which respond (for instance) only to a line of a certain orientation, or an edge moving in a particular direction. Indeed, it was the computational arguments about perception put forward by *Pandemonium*'s programmer which first suggested that feature-detectors might exist. This was what prompted neurophysiologists to search for them (first in the frog's retina, later in the mammalian brain).

This is just one example showing that computational models of psychologcial processes may help neuroscience. Computer scientists can tell neuroscientists nothing about the material nature of the brain; but they may (as in this case) be able to suggest what sort of functional unit neuroscientists might fruitfully look for. Both approaches are necessary: the neural matter is of interest to us primarily because of the psychological functions it supports.

It was precisely because the computational functions necessary for sight had not been properly identified, that these early parallel-processing models of vision had to be abandoned.

Despite their apparently brain-like organization, systems

such as *Pandemonium* were radically incapable of seeing (describing) solid objects – like cubes, pyramids, or teddy bears. They were fairly successful at recognizing simple patterns (such as letters or doodles) – provided these were of a certain size, presented alone, oriented the right way up, and centred in the visual field. But they could not interpret patterns *as* two-dimensional representations of three-dimensional objects.

The reason is that they knew nothing about the third dimension, nor about how three dimensions can be projected into two. They responded to patterns as mere patterns. This suffices for the recognition of alphanumeric characters: there is no need to move into the third dimension to recognize 'A' or 'B'; but for ordinary vision, it does not.

What seeing creatures need to see are things in the real world – food, predators, and pathways. Admittedly, mere patterns are sometimes biologically important. For instance, the red spot on a chick's bill may be the trigger that leads the mother-bird to feed it; and superficial patterns on fur or feathers are often crucial in releasing courtship behaviour. In general, however, vision is not pattern-recognition, but image-interpretation.

It follows that the mere counting of pattern properties cannot suffice to enable a system to see. As this point was realized, the pendulum in computer vision swung away from computerized neural nets. Research turned instead to visual interpretations, images being viewed (*sic*) not as patterns but as representations of the real world.

These interpretations were justified by projective geometry, which tells us how solid objects of certain types would appear to an observer from different points of view. So systematic geometrical knowledge about 2D-to-3D mapping was built into computer programs for 'scene-analysis' (as opposed to 'pattern-recognition'). These programs used their stored knowledge to build sensible 3D descriptions of objects, given depictions of those objects in 2D line-drawings. A drawing of a cube, for instance, would be recognized *as* a representation of a cube.

In general, a scene-analysis program could interpret line-drawings of those types of object which it knew about already. High-level knowledge about a given class of objects would be used to guide the visual interpretation: thus these programs 'knew what to look for' in the 2D image.

For example, they knew that the corner of a cube may appear in an image as a fork-shaped vertex, an arrow-shaped vertex, or an ell-shaped vertex. Given optimal lighting conditions, so that no edges are obscured, projective geometry guarantees that it simply *cannot* appear in any other way. What is more, an arrow-shaped vertex in the image *can only* correspond to, or represent, a convex real-world corner: one that points towards the eye. (See for yourself: try watching one corner of a cube, gradually turning the cube towards and away from you.)

Since they knew what to look for, these programs could adapt to poor lighting conditions, to some degree. Thus if two adjacent cube corners had been found (depicted by two linked vertices in the image), they might deliberately look in the appropriate parts of the image for lines fainter than would normally be acceptable. This was possible partly because of the serial processing: interpretations of image parts that had already been achieved could be used to influence the interpretation of image parts looked at later.

It is no accident that my main example here has been a cube, rather than the teddy bear mentioned earlier. Teddy bears were, in effect, invisible to scene-analysis computers. Such cuddly toys could not even be identified by these programs as solid objects. And no teddy bear could be described by them as something (albeit an un-nameable something) with a smooth furry surface, having bits sticking out here and there, and two shiny, round bumps near one end.

There were three basic reasons for the invisibility of teddy bears. First, because of the simple projective geometry used by scene-analysis programs, only objects with straight edges (or extremely simple curves) could be described by them.

Second, these systems had to be pre-programmed with detailed knowledge of what they were going to see, if they were to see at all. Since they were not told (and moreover could not be told) what teddy bears look like, they had no way of finding their salient curves or surfaces – no way of seeing them.

Third, scene-analysis programs could not see localized depth or surface texture. Even a fabric-covered cube (such as one might give to a very young baby) could not be described by a scene-analysis program as furry or smooth; and no glass cube could be seen by them as shiny.

The human brain is not so limited. It can interpret images of very unfamiliar objects, inferring a great deal of 3D information about them in the process. If you have never seen a teddy bear before, you will not be able to recognize or name one; but you will be able to describe it in detail, as a specific (unfamiliar) physical object. That is, you can see its shape, surface texture, orientation, size, position, and colour; and you can distinguish certain parts of it, such as its shiny glass eyes sticking out from the main surface.

Computer-vision programs, if they are to be useful to us in a wide range of circumstances, ought to be able to do the same sort of thing.

Some recent computer systems can do so (at least to a significant degree). What's more, they are decidedly 'brain-like' in certain ways. Once again the emphasis is on parallel processing, and multiple interactions within networks of elementary dedicated units. But these latest systems are significantly different from *Pandemonium*.

Their difference is theoretical rather than mechanical: their superiority owes less to modern technology as such (even now, most 'parallel processors' are actually simulated on traditional digital machines) than to our deeper understanding of what it is, and how it is possible, to see something as a solid (3D) object, given only a 2D image. (Scene-analysis workers took such questions seriously, but as we have seen they did so in an insufficiently general way.)

The theory implicit in the new 'brain-like' machines is the physics of image formation; for the function and interconnections of the individual processing units are carefully engineered (and/or programmed) according to detailed optical knowledge. The optics concerned does not merely describe the behaviour of light, considered in itself. Rather, it deals systematically with the ways in which light can be reflected from *physical surfaces* of various sorts.

Nor is theoretical optics limited (as scene-analysis was) to telling us what images could represent the corners of cubes, or the tips of pyramids. It answers much more general questions, such as: how is light reflected from a particular sort of surface? or from a surface (or tiny part of a surface) which happens to be oriented at a particular angle relative to the viewer? or from a surface at such-and-such a distance from the eye (or differing distances from the two eyes)?

In these 'connectionist' machines, each processing unit is dedicated to seeking a specific type of perceptual (3D) interpretation for the (2D) image part it looks at. The image corresponding to the overall surface of any physical object is made up of many tiny areas, or point images. Neighbouring point images are likely to be similar, because neighbouring surface points are usually similar. For example, a furry point is usually surrounded by other furry points, and a glassy point by other glassy points: only round the edge of a glass eye, for instance, will this not be true. Difference boundaries in the image such as this one may therefore be interpreted as real boundaries between distinct objects in the real world.

Of course, in a world where tigers and dalmatians exist we would not want *every* discernible difference boundary in the image to be interpreted in this way. The white and black patches on a dalmatian's coat are not different objects: they are part of one and the same physical surface, attributable to one and the same thing (the dog). In general, local differences in the image may be organized, so as to reflect (*sic*) surface markings (such as spots and stripes), and surface contours (such as the gentle curves of a teddy bear's face).

So, much as *Pandemonium* had several levels of 'demons' (looking for stroke-fragments, letter-strokes, whole letters, and words), these recent visual systems have several levels of description available to interpret the image. Some units look for certain sorts of organization within the descriptions arrived at by lower-level units.

One of the ways in which image points can be described by these new brain-like machines is in terms of their distance from the viewer. This is crucial in deciding where one object starts and another begins. For if a sudden depth-disparity is noticed at a series of neighbouring points in the image, these are taken to correspond to the edge of the physical object concerned (such as the side of the bear's eye, or leg). In consequence, these machines could pick out (as you could, too) a dalmatian dog lying on a black-and-white spotted rug; and they could do this even if they had never seen a dalmatian before. (They would not know it was a dalmatian, but that is a different matter.)

The interconnections between the individual units do not merely enable messages to be passed from lower-level to higher-level units (as in *Pandemonium*). They also allow for feedback between units. This feedback takes into account the physcial possibilities of images in various real-world situations.

This would be helpful, for example, in distinguishing a black image patch caused by a black spot on a dog's coat from an immediately contiguous black image-patch caused by markings on the rug the dog was lying on. Suppose that some early processing units had described this part of the image as 'one' black patch (for that is how it appears, in the image). If the depth-detecting units were then to decide that there was a line of depth disparity running through this area, they could pass messages to the patch-detecting units concerned, so as to inhibit them from describing this as 'one' patch (ascribable to 'one' surface). Conversely, feedback facilitating certain units might occur. In effect, then, these new-fangled demons not only shout more or less loudly on their own account, but tell each other to shout more or less loudly, so as to arrive at a mutually consistent set of shouts.

In short, we have here a parallel-processing system, whose units are analogue rather than digital (for they can shout more or less loudly), and which are dedicated to seeing certain things (the units which can see depth cannot see lines or blobs). That is, we have something significantly like a brain. And at least some of the 2D-to-3D computations done by these systems are arguably like those carried out by our own visual cortex. For instance, stereopsis (depth vision that relies on disparities between the image presented to the two eyes) is much better understood as a result of this recent computer modelling research.

This does not mean that all work in computer vision not based on brain-like machines is a waste of time. The need for detailed knowledge of optics was shown partly by previous work in the very different computational tradition of scene analysis; and there are many problems about vision which remain. For example, maybe once you are familiar with dalmatians you *do not have to* build your dalmatian interpretation 'bottom-up', by using each individual mark on the dog's back. Admittedly, you may have to do something like this (perhaps even consciously) if you are shown a dalmatian lying on a black-and-white rug. But what you *can* do is not necessarily what you usually *do* do. The 'top-down' processing that enables you to use your high-level knowledge to see quickly and efficiently in everyday situations (no confusing rugs) may rely on methods more readily associated with traditional artificial intelligence.

In sum, maybe you do not need a brain to be 'brainy'. But there are strong reasons for believing that any competent visual system – whether engineered or evolved – must have a basic organization broadly similar to the human brain. This is necessary in order to achieve the computational functions necessary for the 2D-to-3D mapping which is essential to vision. But functions are where it's at: protoplasm has nothing, essentially, to do with it.

9

Artefactual Intelligence

Patrick D. Wall and Joan N. Safran

'Intelligence' has come to mean some admirable property of humans and animals which allows them to perceive and conceive. It relates to an ability to detect biologically relevant order and meaning in a world which is full of ordered and disordered events which are not biologically relevant to the intelligent individual. Professor Margaret Boden proposes that machines exist or could exist which share some of these intelligent properties. We propose to examine this claim and to question it.

First, we need to be cautious about the phrase 'could exist'. A Turing machine is a real possibility because it has been exactly defined even though it has not been built. However, one must be wary of promises for some vague future machines in the same way that kings learnt at considerable cost to be suspicious of alchemists who, having extracted lead from ore, promised gold and the secret of eternal life. Second, it should be noted that although the exact nature and mechanisms of our intelligence is still a mystery, this does not mean it is an insoluble mystery. 'Mystery' need only mean ignorance, and this does not demand resort to the mystical.

Is there an analogy between machines and brains?

When Professor Boden writes 'There are strong reasons for believing that *any* competent visual system – whether engineered or evolved – must have a basic organization broadly similar to the human brain', what does she mean by 'similar'? Professor Boden stresses that she does not mean that they must have the same physical constitution, and she is right to underplay the physical substance of brains (membranes and cells) and of machines (transistors and wires) when discussing the requirements for simulating functions. A mousetrap could equally well be made of wood or metal. If the similarity between a mechanical visual system and intelligent perception is not meant to hold on the physical level, we must be in the realm of analogy.

There are at least three types of analogy which could be relevant here. The first is *'poetic analogy'* which is a shorthand way of hinting at a relationship between the unknown and the familiar. For example Rutherford's description of atomic structure as electrons in orbit around a nucleus, like planets around the sun, is purely poetic. He had borrowed the word orbit because no suitable word existed. He did not mean that the nucleus and electrons were like the sun and planets, nor did he mean that the electron's movements were determined by gravity and momentum which govern the planets. To take a machine example closer to Professor Boden's interests, it is a poetic analogy to say that the clock tells the time or that radar sees aircraft. This is shorthand for 'People build a machine which goes round and round so that they can tell the time' or 'People build radar machines as part of their equipment to allow them to see aircraft.' Even a talking clock does not tell the time, in the sense that it does not know what it is talking about. Even if we incorporate in the machine components to improve resolution which are similar in principle to components in our own brains, there remain crucial aspects of the seeing individual which are totally missing from the machine. There are sections of Professor Boden's chapter which make it seem she thinks that if enough competent discriminating circuits were stuffed into the radar it would indeed see: she would be wrong.

A second type of analogy is *'evocative'*. Here a principle from one sphere is transferred to another. We find a very good example in Richard Gregory's book *The Mind in Science* (1981). Until the Middle Ages everything that moved seemed to be obviously pulled or pushed. How then did people explain the movement of sun, moon and stars? The answer was to postulate that chariots of fire pulled the sun and that heavenly timekeepers maintained the planets on course. Clocks were invented and it became apparent that if the friction of the axle bearings was reduced, the operation of the clock was prolonged. A completely frictionless clock would run perpetually. If the planets were in a frictionless medium wouldn't they also go on and on since there was no need, after their creation, to maintain the push and shove of a heavenly timekeeper? Here was a constructive transfer of principle from one sphere to another where a particular observation was extrapolated in a legitimate way to form an alternate testable hypothesis. There was no suggestion that planets were clocks or even organized like clocks. The suggestion was only that one aspect of planets might operate on the same principle or law as one aspect of clocks.

The most interesting type of analogy starts as a hypothesis and becomes a statement of *structural or organizational identity* on further investigation. Harvey (1578–1657), who discovered that blood circulated, suggested that the heart was like a pump. He lived at a time when von Guericke (1602–86) had developed pumps with flap valves and with filling and emptying phases. Further investigation shows that the heart is not *like* a pump, it *is* a pump structurally, even though it shares no molecules in common with an engineered pump and its power source differs from all manufactured pumps. Similarly, Galvani in 1771 and Volta in 1792 had shown that electricity applied to a nerve produced muscle contraction and Walsh (1773) had shown that the electric fish, *Torpedo*, actually produced electricity. This led to speculative analogies on natural, human-made and animal electricity. We now know that these are identical in basic organization even though the charged particles and the batteries are fundamentally different.

Of the three types of analogy – poetic, evocative and organizational identity – Professor Boden proposes an organizational analogy between mechanical visual systems and intelligent perception. We wish to show that the analogy is at best poetic or evocative and pretty weak at that. This is not because mechanical visual systems do not or could not have a basic organization broadly similar to that of the human brain. They do not, but they could. It is because mechanical visual systems are not *competent*, i.e. intelligent, whilst most creatures who have brains are.

An historical excursion

There has been an historical obsession to explain complex things which are not understood, in terms of simple things which were thought to be understood. Starting as poetic or evocative analogies, these so-called explanations were confused and quickly assumed to be statements of complete and organizational identity. As science and technology advanced, each generation was presented with new sources of analogies. The physicists of Galen's day 'knew' that earth, air, fire and water were the elements, and therefore Galen constructed a theory of disease based on the imbalance of these four elements. In Descartes's day, hydraulics was the new technology and so the brain was run by hydraulics. By the next century animal magnetism was the thing and by its end electricity had become the life force. For Sherrington, the brain was 'the enchanted loom'. The business of science and technology in this century has generated a superb smorgasbord from which the analogy-makers can pick. Thermodynamics, information theory, Heisenberg's indeterminacy and Bohr's complementarity have been used to explain everything from fetal development to God. Cybernetics was selected by sociologists and economists to generate a mountain of hogwash purporting to explain society and its economics. For our parents the most complicated piece of communications equipment they knew was the telephone exchange and so the brain was like a telephone exchange. For our generation it is a computer. We are re-enacting a historical play in modern dress.

Machines which calculate have a long history, and so do machines which control. By the early 1950s cybernetics formalized the design of machines which incorporate a goal and a means of approaching that goal. Practical artificial intelligence grew from cybernetics with the aim of superseding some unacceptable human qualities. Both the military and industry reacted against their experience of people as slow, expensive, forgetful, unreliable, inaccurate and querulous beings. A person was at best 100 kilograms of mainframe consuming 100 watts of power and operating only once every 100 milliseconds. The major attack point of the past 30 years has been to produce machines which would translate one specific design into a particular finished product. They have failed completely to mimic the intelligent selection of tactics of the skilled machinist who can sculpt a block of metal to achieve the end-result specified by the design. The reason for the failure is that there is an infinite number of tracks leading from blueprint to the object it represents. We can solve anagrams, play chess and mill a cylinder head from a block of metal better than machines because our intelligence not only detects origin and goal but also selects optimal trajectories between the two in the light of properties of intelligence not shared with machines which we will now discuss.

On 'intelligence' and other terms

Professor Boden directs her attention to how machines and brains are organized, and not simply what they do. It is of course correct that functional equivalence implies nothing whatsoever about basic organization. A mouse is equally well killed by poison or a trap or a mouser. However, the claim is made that machines have become significantly like intelligent perceivers because they have recently had incorporated within them certain types of organization which we now know are included in brains. She emphasizes particularly parallel processing rather than the old machine serial methods, analogue versus digital computation and the use of dedicated cell systems. We ask whether even such new generation machines can be considered intelligent.

Intelligence is not just a value term as used by Professor Boden. 'Image interpretation' is not simply having dedicated cells, nor does it boil down to the transformation of a projection from two or three dimensions. 'Seeing' is not adequately defined by the ability to produce 'accurate descriptions'. Being 'brainy' is not just computational functioning. When you borrow a concept, you borrow the theory in which it is embedded. While computer researchers have simulated simple aspects of the brain, they have not shown that further elaboration will generate the theory in which the concepts are embedded. We will argue that intelligence cannot be divorced from strategies organized by purposes, that interpretation cannot be understood at all without meaning, that seeing on any level presupposes theory, and that being 'brainy' is not a property of isolated brains. The artificial intelligence definitions in all these cases lack reference to the significant properties of intelligent beings even if they do avoid the 'difficulties about consciousness' by not discussing it.

The present computer revolution does not aim at intelligence

First we need to clear the decks of what none of us are talking about. A computer revolution is in progress which is changing the world. It depends on machine ability to handle precisely defined tasks with enormous speed, capacity and reliability. Marvellous as these machines are, their properties are the antithesis of any normal definition of intelligence. Let us take the example of aircraft navigation, a spectacular achievement in spite of the curious behaviour of Korean Airlines' 007 which flew a straight line from Anchorage towards Seoul instead of the normal dog-leg. Strong suspicions exist that intelligence overrode the machine. The new navigation is the old navigation in principle improved by the invention of four new pieces of hardware. The crystal oscillator replacing the chronometer, which had itself revolutionized eighteenth-century navigation, increased time precision by a million-fold. The gyroscope replaced the compass and the sun and the stars as stable objects to which one could refer. Memory devices replaced the book of tables and instructions allowing rapid access. Above all, the transistor allowed for the possibility of not only rapid but reliable switching. Second World War general-purpose computers, such as ENIAC, which were used daily to crack the German Enigma code, employed the same principles as modern computers but contained vacuum tube switches. These valves used a lot of power and were bulky but the crucial factor against them was that they broke down a thousand times more rapidly than transistors. This predictable breakdown limited the maximal size of a computer since it was easy to calculate that a size would be quickly reached where the machine would never work, because somewhere in it some vital component would be broken. The transistor has allowed a very large expansion of the size of a computer (*pace*, the readers' personal experience of inoperative machines). This combination of these four components, plus older technology, plus clever design, allows a precision of navigation far beyond the ability of any team of human navigators. The course is simultaneously and instantaneously calculated from inertial navigation data and cross-checked with radio beacons. Marvellous but not intelligent.

The intelligent animal not only detects order in a disordered world but detects that order which has a biological meaning in terms of survival. Furthermore it devises alternative strategies to move from the observed present to the desired future. The navigation machine was invented to serve an exact human purpose. Given its programme, the machine cannot generate alternative strategies for achieving the goal – such as hitching a ride on a passing ship as does a migrating bird, or, as the faithful will do on Doomsday, spinning in their graves to dig an underground tunnel straight to Jersusalem – nor do we want it to.

Alternative strategies in context

An essential difference between us and artificial intelligence machines is that we operate in fundamentally different contexts from the machine. Machines were kicked off by people who flipped on the power. We animals have a history. We are organisms which are the antithesis of general-purpose computers in that we have evolved as a result of steady investigation into a vital problem: how to survive. We living things, all on our own, have come up with a huge number of different solutions to the problem. The number of solutions is equal to or larger than the number of extant species. The variety of solutions immediately points to the extraordinarily anthropocentric narrowness of Professor Boden's strong statement that '*any* competent visual system – whether engineered or evolved – must have a basic organization broadly similar to the human brain'. Tell that to an owl, a bat, a dolphin, a pit viper or a mouse and, in the case of the owl, it will hoot. All of them have solved the problem of seeing in the dark; they can see in the sense that they can detect, classify, identify, locate, plot and predict the course of objects at a distance. We do not know how they do it, but we know enough to be certain that it is not with a 'basic organization broadly similar to the human brain' unless one uses 'broadly' to be so broad as to be trivial and useless. The bat sees through sound. Its brain is organized to measure distance by time intervals, to measure vectors by frequency modulation and so on. In other words, it sees with a basic organization in no way similar to the human brain. It seems that human brain-likeness is not necessary for all competently performing visual systems – certainly not for evolved ones. These animals who would not be graced with the accolade of intelligence by our intellectual friends have evolved solutions which the 'intelligent' machine cannot, and has been programmed not to, contemplate. Our existence as a species is proof of our historical problem-solving success.

Selective attention in context

In addition to, and closely related to our evolutionary historical achievement, we have the ability to learn and remember our general cultural and ethical heritage and our personal individual experience. This is no random passive encyclopaedic absorption of all the events which fire our sense organs. It is clearly selected by some process we call attention, some inherited and some learnt. We know almost nothing of the process by which this selective attention operates, but such 'image interpretation' goes well beyond what could be achieved by dedicated cells or feature detectors. Attention is clearly crucial in forming the brain processes by which we analyze present events and decide on future action. It is a likely guess that this selectivity is based on, and extends the evolutionary process of, natural selection. In other words, just as evolution seems to resolve a particular set of problems for each species, selective attention also implies an attempt to resolve restricted problems.

A measure of our ignorance of the process of selective attention is our lack of understanding of its fragility and its ability to enter what we please to call pathological states where unusual selections are labelled neurotic or psychotic.

We pay attention to events in at least two contexts. One context is that we perceive an event in the light of our knowledge about the entire present world of our senses, about the specific situation and about our historical past. The other context is that we perceive an event in the light of how we should react to it – in the context of its biological meanings.

The first 'environmental' context is just beginning to be fed into machines in an astonishingly crude way. It is impossible to tell a machine how to do it in the sophisticated way that it is obvious to even a gerbil. Packing a machine with information is not the same as the animal's collection of data in context with purposeful exploration to obtain more data and with selective attention.

The second context, the context of meaning, is anathema to the machine intelligence enthusiasts as it cannot be made part of the instructions to a machine. The problem with general-purpose machines is not so much that they lack specialized cells but that they are formal systems. Whether digital or analogue, 'they work by manipulating intrinsically meaning-less formal symbols, whose meaning is ascribed to the machine by the human programmer and/or user'. This feature, once articulated by Professor Boden, is promptly dropped in favour of the 'dedication' point, presumably because it cannot be met even by the newer bigger and better machines.

There is a difference in kind between *any* process of a computer and *all* intelligent processes of animals. Symbols, after all, are by themselves just objects in the world like any other object. They are marks on paper, sound waves, street signs, and so on. Representation in the sense of taking an object as a symbol is not the same as representing a two-dimensional image more 'realistically' as three-dimensional. Taking an object to represent something is of a completely different order from constructing some geometrical 'resem-blance' of an object in the world. To say that a machine performed a 2D-to-3D transformation of a cube is not to say that it recognized the 3D depiction as a representation of a cube: it does not view this depiction as representing some-thing in the world.

If the study of artificial intelligence is the study of machines that manipulate symbols which are *not* symbols to those machines, artificial intelligence emphasizes the artificial and loses the intelligence. The most artificial aspect of artificial intelligence is its lack of real (historical and meaningful) context.

On seeing

For Professor Boden, 'seeing' means the ability to produce accurate descriptions. It is not clear what form these accurate descriptions should take as sometimes they seem to consist in reports about the physical characteristics of *objects* (e.g. '. . . it's about a foot long, with an undulating spotted surface slanting away from the ground'), and at other times they seem to consist in reports about *image* patches (e.g. giving the size and location of black-and-white image points).

It is also not clear at what level the ability to produce accurate descriptions applies. Sometimes for Professor Boden the description can be *shown* by 'appropriate physical behaviour'. If this is her criterion for success we need a 'Foundation for crippled computers' to care for the most advanced machines yet produced.

Another level of accurate description is saying that you have seen a dalmatian. This may be said while the perceiver may be unable to describe a single spot on the dalmatian accurately. It is apparent that the ability to describe spots is not necessary to perceive a dalmatian. If we were asked to give evidence of our claim that we see a dalmatian by building it up spot by spot, we would probably make quite a pitiful mess of it as compared with a computer, which would neatly arrange the spots without ever spotting the dalmatian.

At an unconscious 'basic organizational' level, the physiology of our visual system may well accurately 'take in' image patches and engage in 'connectionist' machine computations to build them up into objects. This would not necessarily enable us to produce such descriptions by taking appropriate action or by speaking in English, nor is such a 'dissectionist' analysis sufficient to account for our ability to behave relevantly towards objects. The mere existence of an algorithm does not constitute an explanation even if an algorithm existed, which is doubtful.

Firstly, seeing cannot be understood as 'the ability reliably to describe things in the external world, given information provided by a visual array'. This is an empiricist account of seeing where all the interesting questions lie buried in the word 'information'. Psychologists have left such a view far behind. Secondly, analysing the seeing of a dalmatian in terms of building 'your dalmatian interpretation "bottom up" by using each individual mark on the dog's back' sounds dangerously close to a reductionism long abandoned by psychologists. Analysing perceptions into their basic atoms is a proper function for physiologists, but these scientists are not so presumptuous as to claim that their discoveries explain perception; only that they provide data on which perception may operate.

What do neuroscientists know about the brain?

Precious little. The last great frontier is between your ears. There lies an astonishing land, wide open for exploration and surely full of immense surprises. Neuroscientists are an intelligent and busy bunch, but they are just beginning to nibble at the beginnings of a few of the tractable problems and must postpone consideration of the great mass of fascinating theoretical and philosophical problems. If we take pain as an example of our knowledge, we can see immediately the depth of our ignorance. Pain to the sufferer seems to be a unique sensation triggered by injury. It has been treated as such by the information theoreticians who take it to be created by a single decision by the brain which has detected the relevant signal. Supposing we place on one side for future study such factors as suffering, emotions, and memory and simply ask what is the signal which announces injury? The fact is we do not know the code for injury, the structures which transmit it, nor the context in which it could be 'received'. It may therefore be considered a trifle premature to start programming a machine to simulate brain functions when we are so ignorant about what it is we are to simulate, both in detail and in general principle. There is no doubt that the neurosciences are making tremendous advances when compared with our previous ignorance, but it is minuscule when compared with what remains to be understood.

The rapid relative increase of basic knowledge has posed continuous problems for the artificial intelligence community. In addition, four exceedingly brilliant men – Pitts, Craik, Turing and Marr – all died at tragically young ages when the real power of their contributions was only just beginning. Their task, and that of their colleagues, was to generate what can be summarized by the title of the seminal 1943 paper by Pitts and McCulloch: 'A logical calculus of the ideas immanent in nervous activity'. That paper consisted of an analysis of its first sentence: 'Because of the "all-or-none" character of nervous activity, neural events and the relations among them can be treated by means of propositional logic.' The paper had an enormous impact, particularly on the growing digital computer industry – but it also gave rise to problems. The first problem was that Pitts and McCulloch themselves rapidly abandoned the first half of their key sentence and moved from a digital all-or-none system to one operating on analogue signals, as Professor Boden does. They moved because their own investigations of the real nervous system forced them to.

The second problem involves 'neural events'. If we come to the present and read the posthumous 1984 book by Marr, *Vision: a computational investigation into the human representation and processing of visual information* (Freeman, New York), it contains a brilliant synthesis of how the components of the nervous system known at the time might operate and co-operate to generate certain aspects of vision. The problem is that, since he began his synthesis, the work of the very scientists on whom he relied to provide him with the raw data has moved ahead to a level where the components he uses as fact are no longer recognized.

The third problem related to 'propositional logic'. The language of propositional logic is not only formal in that it consists of uninterpreted symbols, but it is also truth-functional. This requires that the truth-value of compound formulae be uniquely determined by the truth values of its simple parts and the definition of their connectives. A natural language such as English, or the sequential flow of an animal's movements, is not truth-functional. The result is that a computer using an artificial language based on the language of propositional logic is unable to express explicitly what fairly moronic (but intelligent) speakers of English can. Propositional logic cannot adequately express quite basic statements such as ones involving 'because'. The truth values of the parts do not uniquely determine a truth-value for the whole. In English two true parts connected by a conditional can result in a true or a false compound statement, depending on whether or not there is a causal connection between the two events described. Not too surprisingly, another type of statement which eludes adequate formulation in the language of propositional logic are statements such as 'John believes that the sun will rise tomorrow', which does not break neatly into two simple statements ('belief' seems to be more a relation than a connective), and the truth-value of this statement varies only with the truth or falsity of whether John really has this belief (the truth or falsity of the belief itself is irrelevant).

In summary, because of the delay inevitable in the difficult task of synthesis of components into circuits and the need for simplification, the results in artificial intelligence research trail reality in crucial respects.

Could artificial intelligence improve?

There is no doubt that the very able group involved in artificial intelligence work will generate some extremely interesting pure and applied mathematics aided by the impressive new generation of computers whose design they have influenced. They will be well funded by the military and by industry who are bemused and confused by their own propaganda. We must batten down the hatches to prepare for the storm of promises from the salesmen offering smarter 'smart' bombs and cuter 'cute' robots. From this aspect we will be lucky if we get lead, let alone the promised gold.

What of the future co-operation between the brain studies and machine studies? So far the main contribution from artificial intelligence to psychology and philosophy has been the *problem* of the alleged analogy between artificial brains and intelligence. They have given an aura of respectability to the issue but have contributed little in the way of empirical or theoretical advance. The statement made by Professor Boden that the programmer of *Pandemonium* 'prompted neurophysiologists to search for . . . feature detectors' is not true. One of us was a close witness of these exciting but independent efforts. Slowly, artificial intelligence workers have moved from digital to analogue, from serial to parallel, and finally have just recognized that the object they seek to discriminate exists in a world of which the machine must be aware even if it is only the table-top on which their beloved blocks rest. These exercises may at best generate evocative analogies about discriminative components in our brains, and may even share some principles in common with intelligence; but what hope for simulating intelligence on the basis of basic organizational similarity with the brain?

Two related aspects separate the intelligent perception of all mice and humans from the performance of machines.

(1) Intelligent perception always takes place in the context of both immediate and historical perspective, and involves selective attention.
(2) Intelligent perception always takes place in the context of goals and actions and involves meaning expressed in terms of the biological needs of the whole organism.

On the first point, selective attention is a dominating aspect of our mental activity and living creatures show clear signs of the selectivity. Our sense organs are continuously bombarded with a barrage of data about the details of the world around us. Somehow we continuously reject 99.9% as irrelevant and select out those features which are of importance to our problems of the moment. If there is one mark of intelligence it is this selective ability. Faced with a new problem, goldfish, gophers and gorillas bumble around exploring their environment, try out all sorts of activities and suddenly 'the penny drops'. From that time on they ignore the huge mass of available sensory data and concentrate on the tiny relevant fraction. How they do it is a still mystery, and there is therefore no way at present of teaching a machine how to do it. Of course, with hindsight we can sometimes tell what are the relevant clues and then there is no problem in teaching them to a machine or a student. Hindsight is an example of directing selective attention in historical context. We can order a radar to ignore stationary objects or those with flapping wings, but that is our intelligence not the machines'.

On the second point, the trees on the horizon are not just perceived by us as trees, yes or no. At the same time we perceive them as a bad place to shelter in a thunderstorm, a likely place in which to look for a squirrel's nest, a good place for mushrooms, a bloody nuisance when you want to reap the wheat, a possible source of firewood, healthy, diseased, spooky. In this respect you will perceive the tree in a different way if you are a mouse or a woodpecker or a particular person. You perceive the tree in the context of how you understand your needs and goals, its limits and potentials, and relating these two (you and it) through future activity. In short, you see a tree in terms of its meaning for you. This characterization of perception has not been considered by the designers of intelligent machines partly because no concept of meaning is inherent in any proposed intelligent machine. It is retained by the machines' creators. Partly, too, this is because of the intellectual position of artificial intelligence experts – the Locke, Berkeley and Hume tradition of the isolated insulated purity of objects independent of their relation to the observer.

Meaning is related to the whole organism and its environment, not just to the brain. No amount of similarity in basic organization to the human brain will suffice to model meaning.

Can the artificial intelligence enthusiasts set themselves up as creators of intelligence which would incorporate these factors? Could machines be developed with the selective attention needed to solve historical problems and with the ability to comprehend the meanings required for life? These are not properties of formal systems. All that could be done is to model brains in their abstract (non-physical) organizational structure.

The real question is begged by Professor Boden's title: 'Does artificial intelligence need artificial brains?' This question ignores the issue of the correctness of her description of the machines and the field which studies them as artificial *intelligence*. We have questioned the assumption that being 'brainy' is purely a matter of the brain. Instead we have tried to show that being 'brainy' is a matter of living organisms operating in present and historical and meaningful contexts. These contexts may be irrelevant to the abstract organization of our brains but what follows from this is that only the abstract organization of brains could be modelled by formal systems, not that the brain itself is intelligent. Even if the brain is successfully modelled, the artificial formal duplication would not *be* a brain any more than a pump is a heart. More importantly, while the heart is a pump, intelligence is not a brain.

Professor Boden's title is either a tautology or involves a contradiction. If the only thing being modelled by formal systems is the abstract basic organization of the brain (not intelligence), the field should not be called 'artificial intelligence' but 'artificial brain' modelling. Then the title asks whether artificial brain modelling needs artificial brains, and the answer is yes, trivially. If the phrase 'artificial intelligence' is unpacked it is contradictory. We have argued that in so far as something is artificial it is not intelligent, and in so far as it is intelligent it is not artificial. From a contradiction anything follows.

As brain modelling, the field of artificial brain research could provide an organizational analogy for brain studies. But with regard to understanding intelligence, we could better direct energy to the study of those intelligent creatures who build magnificent machines yet hide the mystery of how they do it.

10

Minds, Machines and Meaning

Richard Gregory

Margaret Boden, Patrick Wall and Joan Safran approach the questions of brain, machine, intelligence and mind from different backgrounds, and they have somewhat different commitments. Although Margaret Boden has worked in the biological science of psychology she is primarily a philosopher, specializing in artificial intelligence. Patrick Wall is a neurobiologist, his special concern being with pain and its alleviation. So he is intensely concerned with consciousness; though for the purposes of this debate problems of consciousness – awareness of pain, or red or green, or guilt or intention or whatever – have been put aside. Thus Margaret Boden deliberately ignores aesthetic considerations, observing that difficulties about consciousness 'obscure the main point being discussed'; but what is the main point? It is the status of *meaning*. One might well ask how meaning can be discussed without reference to consciousness. It can, for although consciousness is uniquely 'private', understanding and meaning can be recognized objectively. Margaret Boden and Patrick Wall agree that seeing relevant appropriate meanings in symbols or situations is a major characteristic of intelligence – and they agree that this is remarkably present in a great variety of living organisms and sadly lacking in computer-based so-called intelligent machines. Where it does seem to be present, if in rudimentary forms, Wall and Safran at least would say that it is present in the machine (such as a navigating system) becase *we* put it there. They deny that a human-made machine can appreciate *context*, *needs*, or have selective attention. In this they agree with the American philosopher, John Searle in his recent BBC Reith Lectures. They argue that as computers work by, and are limited to, *formal systems*, they cannot understand symbols (or indeed anything else either), though they can manipulate symbols, according to formal rules, with consummate speed and accuracy far surpassing our own fumbling efforts. Though pocket calculators are extremely useful they do not understand the questions they are asked or the answers they provide. At least for the present this is so also for the largest and most sophisticated computers – *they do not understand.*

But – and this is the key question – will this *always* be true: will human-made computer-based machines *never ever* understand *meanings* of signals or symbols? Those who hold that machines will always be dumb generally argue that there is some special *substance* in biological brains which allows them understanding and consciousness. Those who take the opposite and, it must be said more daring view, hold that understanding, and also perhaps consciousness, are somehow given by *functions* which are carried out by physical brain processes. It is then suggested that these or equivalent functions might be carried out by quite other physical means – such as by silicon chips. Then a computer might not only respond to the meanings of symbols but might even be conscious. This is not to say that it will share our fears or interests, for it will not have our biological backgrounds or physiological restraints or benefits; but, it is suggested, it may in its own way be intelligent. This of course reflects what we mean by 'intelligent'; but it is a pity to limit the word's meaning to that which is just-as-we-are. In any case there is (fortunately) a vast variety of human abilities and, we may suppose, kinds of intelligence. Very rightly Patrick Wall stresses the special biological qualities of animals. He also objects to Margaret Boden's account of perception as 'image interpretation' – urging that this is a quite inadequate description of what it is to see, to hear, to touch and generally to understand by the senses.

The central theme of Margaret Boden's talk is that there is 'no *special* physical property that is essential to intelligence'.

Though, perhaps, as diamond is the hardest of substances, so it might be that, contingently, brain-substance is uniquely able to carry out the necessary processes for intelligence. Or more limited, conceivably only the biological constitution of brains is capable of solving survival problems in real time: possibly human-made components cannot, for example, be wired in sufficient parallel profusion to solve real-life problems in real-life time. But even if artificial intelligence turns out to be more like the slow intelligence of philosophy, mulling over questions for thousands of years, it might yet count as 'intelligent' though it is far from being biological.

Personally I would accept a machine as intelligent if it can cope appropriately with novelty. I would accept that a word processor understands English if it can spell. For this it must appreciate *context*. For example if, reading these papers, it spelled out: 'Eye will put them write' – I might suspect it did not appreciate meaning contexts for 'I' or 'eye', or 'right' and 'write'. Or would it simply be poor at spelling? We have just this kind of uncertainty with children, and even sometimes with our colleagues. I do know, though, that Margaret Boden, Pat Wall and Joan Safran understand so much that it would be invidious of me to attempt to say more: – ore trie too putt there eyedeas rite.

Part IV Towards A New Science

11

Health for All by the Year 2000?

Alwyn Smith

The announcement by the World Health Organization (WHO) of its programme entitled 'Health for All by the Year 2000?' initiated a series of debates which have ramified into a number of aspects the feasibility of the implied objective. WHO itself has been entirely realistic about what it hopes to accomplish, and meticulous in its assessment of what resources will need to be allocated and deployed. Nevertheless, the assumptions common in medical circles in the wealthier nations about what resources are needed to guarantee health have provoked a considerable scepticism about the general feasibility of any such objective. It is almost certainly reasonable to doubt the capability of the world's governments to emulate the health care provisions of the more-favoured parts of the most-favoured countries. The key questions, therefore, concern the necessity of such provisions and the degree to which health is dependent on them. These questions arouse considerable controversy but are of increasing importance to even the rich countries, and are vital for the larger world.

The maintenance and restoration of health represents a substantial pre-occupation of any industrialized society and typically consumes between 5 and 10 per cent of gross national product – depending on how thoroughly costs are defined and identified. Political preoccupations tend presently to be with how this resource consumption should be distributed between public and private sectors, and with how the benefits purchased should be distributed within the population. Relatively little serious consideration has so far been given to the identification of objectives, and therefore little serious thought has been possible as to how suitable objectives might most profitably be pursued.

There is a general tacit assumption that the traditional entrepreneurial practice of medicine will ensure that resources are distributed in accordance with perceived need, and that the only management that a health service requires is basic provisioning and a modest degree of intervention in the interests of ensuring that serious illness does not go untreated because of individual inability to pay for the treatment that is required. Prevention is often held to be either already so well established as to require little continued consideration or so largely a matter of personal behaviour that it is inappropriate for either doctors or the health services to engage in preventive activity. In short, medicine and the health service are expected to be largely preoccupied with the diagnosis and treatment of illness in individuals, who in turn are expected to be the main initiators of transactions with the services.

This somewhat Panglossian assessment of the role and effect of medicine and the health services is so well established that to call it in question is not only unpopular but also viewed as sufficiently perverse as to lay upon those who do challenge it the onus of cogent deployment of argument and evidence.

The impact of medicine on health

The history of health in the last 150 years has been popularly viewed as a record of continuing improvement, and the contribution of medicine to this happy state of affairs is commonly believed to have been substantial. In consequence, the present is usually viewed with satisfaction and the future looked forward to in the confident expectation that the advance of medical science will eventually permit us to prevent most of the ill-health that remains and to cure whatever we do not learn to prevent. Furthermore, there has been a widely held assumption that each part of the world will inevitably pass through similar experiences, and that consequently the recent past history of the most developed countries provides a reliable prediction of the future for the less developed.

This happy analysis rested on a general awareness that, for most of the population of the industrialized countries, mortality rates have diminished substantially and that the average duration of human life is consequently much longer than it was. At the same time there has been a spectacular growth of scientific knowledge concerning the functioning of the body in health and sickness, and a considerable increase in our understanding of disease causation. To suppose that the improvements in the usual duration of human survival and the growth of medical science are related is a natural assumption and one which is still very widely accepted.

However, in the past 20 years or so these simple assumptions have been challenged with increasing cogency. The challenges have been based on examination of the timing of the mortality changes in relation to the growth of medical knowledge, the detailed nature of that knowledge in relation to the changes in health experience, and the degree to which the practice of medicine might be said to have brought the new medical knowledge to bear on a substantial proportion of the sickness experience of members of the population. It is now clear that the onset of the mortality decline considerably antedated both the development of medical science and its effective application to the medical care of the population; that the main diseases for which mortality has declined are those in which effective preventive or therapeutic procedures were not available until long after most of the decline had taken place; and that the virtual cessation of the decline in mortality has occurred at the same time as the attainment of something approaching universal accessibility of scientifically based medical care.

To the medical profession these new assessments have seemed unattractive and unconvincing. Doctors are inclined to assess their personal experiences rather than their collective impact, and most doctors are impressed with their capacity to intervene so as to alter the course of their patients' illnesses and to manipulate biological processes in the light of an apparently well-structured understanding of human physiology and pathology. Moreover, most doctors are continually sustained by the expressed satisfaction of grateful patients as well as by the considerable intellectual satisfaction that is the reward of making a verifiable diagnosis. Since the processes of policy and decision-making in the health care field are almost pre-empted by the freedom of clinical judgement which doctors are almost universally accorded, the new analyses have so far had little practical effect.

The trends of mortality

No one any longer doubts that mortality rates (the risk of dying per unit time) began to fall in Western Europe at some time during the eighteenth century. They fell first and have fallen most for people in the age-range 5–45 and, within this range, the fall has been sustained until very recently when mortality rates reached levels below which there is relatively little room for further fall. A typical British 5-year-old male has a 95 per cent chance of reaching age 45, and for a female the chance is almost 97 per cent at current age-specific mortality rates.

At ages below 5 years, mortality remained relatively stable until much later. It began to fall to any marked extent only during the present century. Mortality at very young ages has remained more intractable and the position in this country now is that perinatal deaths outnumber those in the next 35 years of life.

Mortality after the age of 45 has fallen least – especially for males. In the past 100 years the average duration of life beyond age 45 has increased by less than 5 years for males and the chance of a 45-year-old male reaching the age of retirement (65 years) is only about 72 per cent. After the age of 65 years the position has changed very little since statistics became available; further expectation for males has increased by just over 1 year to about 12 years, and for females by about 4 years to 16 years.

The trends can be summarized by the observation that mortality began to fall first and has fallen most among younger adults; for children the fall began later, and was least marked at the youngest ages where mortality remains substantial; for older adults it has fallen least. The pattern suggests that the causes of mortality in very early life and in later life have been much more intractable than among young adults. This makes sense in biological terms, and in terms of what we know about disease aetiology.

An explanation of the trends

Unfortunately, there is always likely to be some controversy about the detailed reasons for the mortality changes. When they began there were no useful statistics of any kind and during most of the period of major change the diagnostic information on death certificates must be regarded as suspect. Nevertheless, the data have now been intensively studied by a large number of scholars, and there seems no serious objection to the broad conclusion that the principal change in causes of death has been a major decline in death from infections. The epidemiologist Thomas McKeown (1976) provides evidence that about 85 per cent of the total decline in mortality since 1700 is attributable to reduced mortality from infections, and more than 70 per cent of the reduction of mortality during the present century is similarly attributable. He further estimates that more than half of the improvement in mortality from infection took place before the beginning of the present century.

Evidence on the kinds of infection responsible for the change is less reliable and conclusions are therefore less secure. McKeown (1976) provides evidence that, since diagnostic data were available, some 40 per cent of the total decline in mortality has been due to airborne infections, 21 per cent to gastrointestinal infections and 13 per cent to other infections. Of the remainder of the total decline, some 9 per cent has been due to a falling mortality attributed to old age, 6 per cent to prematurity and associated conditions and the rest spread over a wide range of conditions.

The generally accepted view that most of the declining mortality since the eighteenth century was due to a reduction in mortality from infections leaves us with the task of explaining the change in the significance of infection. Such a change might have been due to a reduced incidence of infectious disease (incidence is the risk per unit time of incurring an incident such as an attack of infectious disease) or to a reduction of the fatality rate associated with infections, or to both. In turn, either kind of change might have been attributable to specific measures directed at the infections, or to a change in general circumstances affecting incidence or fatality. In the absence of any useful data on the incidence of infection in the early period of the change we shall probably never be able to reach conclusive decisions on these issues. However, it is possible to conclude that most of the changes occurred before any effective medical measures were available and that such medical measures as were available consisted of preventive procedures implemented by public health authorities rather than treatments administered by individual clinicians. The only significant clinical measure available before the present century was smallpox vaccination; its effect is so specific to smallpox that its contribution to mortality decline must have been very small. In short, most of the decline in mortality had occurred by the time that any specific medical measures were available for general use. Even the effect of chemotherapy and antibiotics must be judged to have been small in their contribution to general mortality improvement, however dramatic their impact on the individual case. They arrived too late on the scene to receive more than a very small fraction of the overall credit.

Controversy continues as to the main reasons for the initiation and continuation of a change in the mortality from infection which began before even the concept of infection was well established – still less its mechanisms understood. If McKeown is right it was the airborne infections that were first involved, and there is no convincing explanation available in terms of climatic or social change favouring a reduction in their spread. Tuberculosis was undoubtedly the most important killer during adolescent and adult life, and we have no grounds for supposing that its human pathogenicity might have changed for reasons connected with either the organism or its antigenicity. Since such phenomena have been studied and understood the organism seems to have been stable.

The most remarkable change that was proceeding in the circumstances of the human species at the relevant time and place was in food supply. Advances in agricultural method and organization, and an increased commercial traffic with other countries, were increasing the available supply of food in Western Europe and lowering its cost. It seems certain that the eighteenth century witnessed a considerable improvement in the quality and stability of the individual's diet and that the improvements have been sustained ever since. McKeown argues this as the most important cause of the decline in mortality, and sees the industrial revolution as supplying its own workforce by its effect on death rates and consequently on population growth. Others have argued the importance of the growth in public education, which was associated not only with rising aspirations but also with the knowledge required for their realization. Evidence from the Third World at the present time supports the view that levels of general education are associated with infant and child survival even where other circumstances are relatively unfavourable.

We are on more secure ground in seeking to detail the changes that have taken place in the past 50 years. These may be summarized as a continued improvement in mortality in infancy and childhood, small improvements in mortality in early adult life and very little change in later adult life.

Most of the reduction in early mortality has continued to be due to a decreasing mortality from infection, and it is at least credible to argue that specific immunizations have been responsible for a part of this, and the use of therapeutic agents in the treatment of infections for much of the remainder. Nevertheless, it is useful to remind ourselves that the provision of a National Health Service, capable in principle of bringing the benefits of scientific preventive and curative medicine to all, has done nothing to change the marked gradient of infant and child mortality across the social class spectrum. It is still true that if we could obtain for all newborn individuals the first-year survival experience of those born to parents from the professional and managerial classes we should save more than 12,000 lives annually in Britain. It is difficult to provide a detailed explanation for this phenomenon – absent from other European countries such as Sweden – but it seems clear that the access of children to good health is still limited in Britain by parental social circumstances. Maternal nutritional history, parental education and housing seem the most likely factors.

The relative modesty of the mortality improvements of the past 50 years in early and middle adult life requires a more complex explanation. First, the spectacular improvements in the preceding century may have left relatively little room for further improvement. Second, such continued improvement as has been possible in mortality from infections and certain other diseases has been offset by increased mortality from non-infectious causes. Outstanding among these have been arteriosclerotic heart disease and malignant neoplasms (cancers).

When we come to consider later adult life, and especially the age period 45–65, the significance of increased mortality from non-infectious disease becomes more evident. The virtual disappearance of infection as a cause of death in this age range, and the apparent diminution in rheumatic heart disease and its long-term complications, has been offset by a substantial increase in mortality attributed to ischaemic heart disease and a smaller increase in respect of malignant neoplasms. The former has shown an increase of some 300 per cent in the period; the latter an increase of 30 per cent.

Opinion differs about how far these changes reflect underlying changes in disease experience and how far they reflect changes in the precision and fashions of diagnosis. Most people believe that there has been an overall increase in mortality at this age period from arteriosclerotic diseases and neoplasms, and that this has offset the decrease in mortality from rheumatic heart disease and chronic infections. However, even within these broad categories the position is complex. In the case of neoplasms, mortality for some anatomical sites has increased, while for other sites it has decreased, while in arteriosclerotic heart disease the trends have been different in the different social classes and among the different industrialized countries.

The explanation of these increases remains far from clear in detail. A most important issue is the sex difference in both mortality and mortality trends. For example, ischaemic heart disease mortality has failed to offset the decrease in mortality from other heart disease in women, so that overall mortality from heart disease has declined slightly for women in middle life whereas for males it has increased by about 50 per cent.

This is perhaps the strongest reason for believing that the increased mortality attributed to ischaemic heart disease is real rather than due to diagnostic transfer.

In the case of malignant neoplasms the picture is complicated by the restriction of some sites to one sex. The most interesting case is cancer of the lung, in which the trend in the two sexes seems similar but separated by some 30–40 years. The evidence continues to accumulate that cancer is largely due to agents in the environment, and that many of the responsible agents are either the products of human manufacture or at least introduced by us into our own environment (Doll and Peto, 1981).

It has become fashionable to refer to the diseases of middle life that have increased in frequency in the past 50 years in the industrialized countries as 'diseases of affluence'. Implicit in this categorization is the notion that they arise from the 'benefits' of industrial activity and speculation about the roles of diet, idleness and sexual behaviour has been rife. It is important to note, however, that most of these so-called diseases of affluence occur disproportionately among the least affluent rather than the more affluent classes in industrial societies, and that although they seem less common in poor (i.e. non-industrialized) countries their incidence does not seem to be associated with per capita income among the industrialized countries. It is possible that they arise from activity involved in the creation of the affluence in industrial countries rather than from its self-indulgent consumption.

Mortality among the elderly has improved very little. This is partly because we are 'born but to die', but also because the effect of the increased mortality from diseases associated with industrialization continues to operate beyond the age of retirement and to offset such gains as have been made in respect of mortality from infections.

For the elderly, the prospect of death is always imminent, and they have not shared in the general reduction in the immediacy of its threat that has occurred among the young. Developments in the treatment of acute infections have deprived the elderly of the relatively peaceful prospect of death from pneumonia and the introduction of investigative and treatment regimes in cardiac and renal failure have added to the problems of dying from these conditions. The growth of chemotherapy in malignant disease threatens new miseries. Only a very young man like Keats could have described dying as 'to cease upon the midnight with no pain'.

Trends in non-fatal sickness

Available information on non-fatal illness is much more difficult to interpret. Most statistics purporting to describe morbidity relate really to service use, and most of the data relate to hospital in-patient care. Admission rates are obviously affected by the availability of beds, and for many diseases admission rates have been affected by changes in the usual duration of stay. This is particularly marked in the case of mental illness where trends in admission rates reflect psychiatric practice much more than disease incidence. We have virtually no idea whether there has been any change in the amount of mental illness, and can only conclude that certain forms of illness – mostly those associated with infections – have become less common.

Statistical reporting of absence from work through illness is very difficult to interpret. Changes in the nature of employment and in provision for sickness absence insurance have probably affected the data more than changes in sickness experience. In most countries the number of spells of sickness absence from work seems to have increased, but there is no way of knowing how far this reflects changes in the nature of work or changes in expectations.

There has been a fairly well-documented reduction in the amount of disability arising from the older forms of pneumoconiosis, and in some industries the rate of accidents has improved. There is evidence, however, that new hazards are constantly arising in the industrial workplace and there is increasing evidence that ill-health may arise from the threat or the experience of unemployment.

Perhaps the most important evidence of changes in non-fatal morbidity is the changed age structure of the population. The increased proportion of elderly people in most industrialized societies makes it certain that the prevalence of the characteristic impairments and disabilities associated with ageing has increased. We have no evidence that social developments will have offset these and reduced the prevalence of the handicaps to which they give rise. A substantial literature has now documented the range and variety of the morbidity associated with an ageing population, and has clearly demonstrated that this morbidity leads to very substantial unmet need.

More elusive, but nevertheless important, sources of change in the prevalence of illness are the changing definitions of sickness and health. It is quite clear that the distinction between sickness and health is often a function of expectations and these are in turn a function both of the vicissitudes of the normal life cycle and of social perspective. If health is defined in terms of capacity to meet legitimate obligations and to enjoy legitimate rewards then changes in the legitimacy of either obligations or rewards may lead to changes in the prevalence of ill-health. Health is sometimes defined as something other than the absence of sickness, but any definition of sickness as the absence of health implies that health and sickness are antithetical states. The distinction between them, and consequently the prevalence of ill-health, may be changed either by unplanned or by planned social change.

The practice of medicine

Medicine is often described as the second-oldest profession, and reference to physicians occurs throughout the records of human existence. The earliest evidence of what might be called medical science stems from the Arab world but the Greeks also developed a system of medicine with a respectable intellectual basis. In Western Europe, however, medical science exerted a tenuous influence on medical practice until relatively recently. It would be difficult to demonstrate that the anatomical and physiological discoveries of the seventeenth century affected the generality of medicial practice and the introduction of such useful drugs as mercury, quinine, ipecacuanha and digitalis was based on empiricism rather than science, and their actual administration was generally not only without effect but also without critical asessment. Among the first scientifically based proposals in modern medicine were those of Percival –

physician in Manchester – who established a programme of public health surveillance and activity based on epidemiological study.

However, during the latter half of the nineteenth century the medicial profession laid the foundations of its professional and scientific development. During this period the distinction between medical science and practice was greatly reduced and the collective observations and researches of medical practitioners and medical scientists developed a substantial body of knowledge and understanding of the origins, processes and remedies available for illness and suffering.

The present century has witnessed further developments but the great growth of what we now think of as clinical science began as recently as the 1930s. The impetus to this development derived partly from the considerable development of medical schools as university faculties, and partly from the growing activity of the pharmaceutical industry. The need to subject new therapies to assessment led to the development of clinical trials which encouraged not only the application of scientific method in the form ot statistically based trial designs and statistical analysis of results, but also the increased use of physiological measurements as indices of outcome.

Since 1930 the possibilities of verifiable diagnosis and specific therapy have increased enormously. The intellectual rewards – especially to the hospital-based doctor with access to the necessary laboratory facilities – have greatly increased. A whole new generation of students has graduated and matured in practice – and even retired – in the scientific tradition. During this time many of the diseases which were formerly common have almost disappeared – rheumatic endocarditis, pulmonary tuberculosis – and have been replaced by others formerly considered relatively rare – myocardial infarction, bronchial carcinoma. But the most significant changes have been due to causes outside the scope of laboratory investigation – they have resulted from the changing age composition of the population and the changing access to medical care. They have brought about an increasing mis-match between the needs of the sick community and the interests and resources of the doctors.

The mis-match was evident first in general practice when

doctors trained in the teaching hospitals were confronted with open access by people whose problems lay largely in the effect of intractable ill-health on their lives and not in their disordered physiology. The response of general practitioners to this situation could be summarized under the headings adaptive, political and defeatist. To their enormous credit the adaptive response has pre-dominated, and the Royal College of General Practitioners has led the whole profession in its assessment of its new role and in the development of new patterns of professional training and practice. General practitioners have also legitimately pursued a better status and reward, and achieved in this respect a notable political success. The relatively few who felt defeated have now either emigrated or retired, and the speciality as a whole is well in tune with the responsibilities it faces.

Hospital specialists have found the problems more difficult. They have often been compelled to admit patients they felt unable to help; they have been criticized for their waiting lists but denied the resources they believe are necessary to reduce them; they are accused of arrogance and indifference if they discharge patients with unsolved social and domestic problems, and of empire-building if they try to run their own follow-up clinics. They are inevitably aware of the negligible influence they can exert on the survival of their most seriously ill patients – those who are very old, or who have cancer or myocardial infarctions. Frequently they see their role as diagnostic arbiter rather than caring physician since the structure of the modern specialist referral system (by no means unique to our NHS) seems to cast them so. Adaptation has been very difficult in most cases, defeatism is alien to a section of the profession which sees itself as an elite that has survived rigorous and exclusive training and qualifying ordeals. Their resort has been largely political: they have contrived to maintain a remarkable freedom of action and judgement, they have cornered a large proportion of available health care resources and they have maintained the possibility of the highest personal extrinsic rewards.

Public health doctors – a numerical minority speciality – have fared least well. Inescapably the direct employees of the public service, they have had to endure a series of unilateral and arbitrary changes in the nature of their duties and the terms and conditions of service. Trapped into consensus management teams they have largely lost their consulting role at individual level. Collectively, their voice as health advisers in the community's affairs has been usurped by the clinical specialities and in any case rarely heard in the clamour of competing advice on health issues now directed at local and central government by a variety of different interests.

For many doctors the last straw is represented by academic analyses which claim that improvements in health have had little to do with medical science or practice, and that doctors should re-orientate both their scientific and their professional perspectives in rather vaguely specified directions.

Nevertheless, it is necessary to say what one believes. As a science McKeown (1979) believes that medicine has placed too much emphasis on the mechanisms of disease processes and not enough on their causes. There has also been insufficient emphasis on their consequences. As a professional practice there has been too much emphasis on diagnosis, not enough on treatment, and far too little on either prevention or care. We have been misled into believing that all problems have technological solutions and have neglected the social and political options as routes to better health. It was once fashionable to argue that politics should be kept out of medicine and medicine out of politics – as if any action capable of having an effect could be non-political. In the nineteenth century one of the founders of physiology and of scientific, materialist medicine, Virchow, said 'Should medicine ever fulful its greater ends, it must enter into the larger political and social life of our times . . .'.

Nevertheless, most doctors spend their working days preparing themselves to offer, and offering, advice to individuals on their health problems. If the advice is to be useful it must not only be soundly based in the theory and practical experience of medicine, it must be relevant to the life and problems of the individual it is designed to help as well as comprehensible and, ideally, persuasive. Anything which places the doctor at a social distance from patients will inevitably impede both relevance and comprehensibility. The problem of maintaining contact with the people one serves while mastering a detailed knowledge that is denied to them and insulating onself protectively, at least to some extent, from their suffering, is one of the most difficult to confront not only doctors but also their teachers.

Implications for the present and future

Since the eighteenth century the average duration of human life in industrialized countries has increased from about 30 years to substantially more than 70 years, but the demographic evidence is clear that this is due to a greater proportion of newborn individuals reaching an 'allotted span' rather than to any appreciable change in the upper limit. It suggests that an upper limit to the length of an individual human life is 'programmed' into our genotype but that in common with other species a substantial proportion of us have always failed to reach it because of exposure to environmental hazards. This analysis suggests that human mortality could have three kinds of cause. First, accidents to the individual genotype which determine an earlier limit; second, environmental insults which override or modify the programe; third, the operation of the basic programme.

Genetic accidents are individually rare but cumulatively quite important, and their importance as causes of early mortality has increased relatively as environmental causes have declined. There is very little present evidence that we shall be able in the near future to avert their occurrence, but reasonable hope that some of them may become detectable at an early prenatal stage so that affected fetuses need not be born. Genetic accidents that are not detectable at an early stage are nevertheless likely to continue to occur. Their victims will often require medical, nursing and other care and support throughout their lives, which may be short but which are increasingly likely to lengthen as therapy and care improve.

Overwhelmingly the most important causes of premature death, as well as both acute and chronic disease and disability, are environmental influences, both known and unknown. Our environment contains both the elements necessary to survival and the agents that hazard that survival. Environmental causes of ill-health include both shortages of sustaining elements and excesses of hazards. The history of mortality in the industrialized countries demonstrates that the environment is capable of alteration, and common sense indicates that it is to a considerable extent amenable to control.

Since the analysis of our past health history demonstrated the crucial importance of favourable environmental change, it had become fashionable during the 1940s and 1950s to underplay the importance of the physical environment in determining health, and to concentrate on behavioural factors with the tacit assumption that these were largely under the individual's control. Health education has been principally directed at individuals, who have been urged to choose more health-conducive lifestyles. Such an approach ignores the significance of environment as a determinant of the relevant 'choices'.

More importantly, it is becoming clear that industrialized life poses new environmental threats whose significance is increased because of their constant newness. The largest single hazard arises from wastes produced by the processing of energy. The combustion of hydrocarbon fuels has already heavily polluted our atmosphere and there is nothing in the newer sources of energy that provides much comfort. The by-products of other industrial processes add daily to the complexity of the pollution problem. In addition to the unintentional pollution of environment by waste there is an increasing problem of pollution by substances deliberately liberated into our environment. Food additives, pesticides, toilet preparations and drugs are intentionally bioactive substances – in many cases intentionally lethal to living organisms – but whose full environmental effects are largely unknown. The concentration in Western Europe of a relatively dense population synthesizing and liberating bioactive substances into the environment must surely be a unique phenomenon in the world's history.

An environmental influence of considerable significance arises from the enormous increase in travel. Although there is no doubt that much human disability and death was caused by horses – mainly from their pollution of the environment by faeces – motor vehicles on land, air and water have been responsible for a unique volume and severity of trauma. The cost and availability of fuels may eventually limit the hazard, but in the shorter term the significance of transport as a source of premature death will increase. Road accidents are already the principal cause of death among young adults in most European countries.

However, the principal source of premature death and disability may well be the industrial workplace. Occupation is, after age, the principal variable that influences mortality. We are not always able to distinguish the hazards of the work itself from those deriving from the lifestyles associated with occupations, but the distinction is not really important. Industrialization has brought a range of new hazards to human existence that are only just beginning to be recognized or understood. In some cases the hazards extend to the spouses and families of workers, for example in lead and asbestos work and in the case of cancer of the cervix which is closely related to husband's occupation.

A new hazard is being recognized as deriving from unemployment. Paradoxically, it seems that although work is often injurious to health, unemployment is also associated with poorer health and higher mortality. The detailed nature of the relationship remains unclear. Societies in which economic depression gives rise to unemployment may be particularly unhealthy for their members. The poorer health of such societies may arise among the unemployed or among those whose continued employment becomes more hazardous as workers' ability to maintain a safe working environment is threatened by fear of unemployment. The problem is newly recognized and poorly resolved.

For the rest, death will eventually supervene from the operation of a programmed limit to the survival potential of our species. The role of medicine in this context is neither preventive nor curative but caring. When we have prevented all that is preventable and cured all that is curable, it follows that what remains is amenable neither to prevention nor to cure.

Medical capability

Much the most striking development in medical practice during the present century has been in diagnosis. The range of human physiological malfunction has been extensively studied

and there now exists a comprehensive and systematic categorization with an elaborate nomenclature. The verifiability of
diagnostic labelling is quite well developed, and at least in
broad terms the diagnostic labels assigned by one doctor will,
in a high proportion of cases, be 'confirmed' by others.

Treatment has been improved in a number of respects.
There is now an extensive range of more or less effective
analgesics – although their safety is variable – and there is a
sufficient number of anti-bacterial agents to justify the hope
that as organisms acquire tolerance to them new agents will be
introduced so as to maintain our present control capacity. A
wide range of infective agents nevertheless remains resistant
to available agents. There are also a number of preparations
available for remedying deficiencies of metabolism, although
the hope engendered by the introduction of insulin treatment
for diabetes has been less than completely fulfilled in relation
to other diseases.

It has to be admitted that there have been no major medical
advances in the treatment of arteriosclerotic or malignant
disease, little of consequence in the treatment of serious
psychiatric disorders, and very little for the treatment of the
extensive and common disorders of connective tissue – a
distressing feature of the medical treatment of joint disease has
been the fatalities associated with what have otherwise
seemed promising introductions.

Surgical treatment has seen some of the more notable
advances. An improved understanding of physiology has
greatly reduced the risks associated with surgical procedures
and facilitated a substantial extension of the range of procedures that can safely be carried out. This has fostered
considerable innovation in intracranial, intrathoracic and
cardiovascular surgery, as well as in the transplanting of
organs from living or dead donors. Perhaps the most signal
advance has been in the surgical treatment of trauma.

Medical preventive techniques particularly include those for
imparting or stimulating an acquired immunity to infections.
A number of previously serious infectious diseases have been
virtually eradicated and more have been substantially reduced
in incidence. For the most part these immunization procedures
have been associated with few unwanted side-effects.

One of the apparently more successful areas of medical practice has been obstetrics. As judged by the survival of the fetus and of the newborn beyond the first year of life the present century has witnessed remarkable improvements in the management of reproductive functions. The causes of the improvements are nevertheless far from clear. Much of the improvement in infant survival is probably attributable to the prevention of infection, and this must be mainly due to improvements in the hygiene of the domestic environment. Improvements in intranatal survival have varied substantially with socioeconomic characteristics of the maternal environment; the principal biological determinants seem to be weight at birth and the presence of congenital abnormalities, both of which also vary with socioeconomic circumstances. There has been some improvement in the survival of very small babies but their frequency in the newborn population still explains much of the variance in perinatal mortality.

In summary

The human species has evolved for at least a million years in broadly its present form, and has acquired in the process a genotype which permits us to survive and even to flourish in a wide variety of environments. Recently, (during the past 10,000 years or less than 1 per cent of our history) we have extended our range of habitable environments and increased the permissible density of our population by a variety of cultural developments of which the most important has been agriculture and animal husbandry. This increased density has aggravated, if not created, the hazards of infection and homicide, compromised the gains from increased food availability and reduced the hazard of predation. In a small part of the world, even more recently, a series of further cultural changes involving a substantial local solution of the problem of food supply led to a significant decline in the hazard of infection, while equally localized changes in fertility permitted the consequent reduction of mortality to lead to a substantial increase in life expectancy and a changed population structure. Meanwhile the associated cultural changes are increasing the hazards of homicide and altering the environment at a rate far beyond any creature's capacity to adapt.

This is the situation we are in. For the first time in our history we actually know something of the range and variety of our species membership and of their problems and predicaments. Nevertheless, we can only guess at what even the near future may bring. We have learned remarkably little so far from our history, and most of us are both ignorant and careless of the necessities of both individual and collective existence.

It is nevertheless possible to spell out the fundamentals. Most human individuals that survive birth have the capacity for useful and rewarding life until they reach the age limits that our evolutionary experience has programmed into our genetic constitution. The imposition of those limits is usually preceded by progressive disability which it is humane to alleviate but fruitless to seek to avert. We have the means to regulate our fertility while satisfying our sexuality, and to prevent infection while enjoying the benefits of substantial population aggregates. There is evidence that we could feed a population of several times our present numbers if we used efficient methods of food production, ate no more than is healthy, and distributed what food there is according to need rather than profit. There seems no reason why we should not contain the liberation into our environment of biologically active substances whose effects on health we do not understand and increase our understanding of the environmental conditions that best support and least threaten our survival. There are grounds for supposing that the amount and nature of the work we need to do to ensure our healthy and happy survival is no more than is consistent with such work remaining both intrinsically rewarding and uninjurious.

The questions that remain are whether we shall achieve these basic conditions for our survival, and how long it will take us if we do. I am not very optimistic in my attitude to either of these questions, although I do not doubt that the species will continue to survive.

12

Organizing for Science

Tom Blundell

In the past 40 years science has undergone a qualitative change in its organization. This has derived from the nature of science itself. Ever more powerful machines are being used to probe smaller sizes, higher energies, lower temperatures and so on, whilst the increasing use of computers and information technology reflect an increasing complexity (Rescher, 1978). Both power- and complexity-intensive developments have led to exponentially increasing costs, which many have seen as leading to a limit of growth, or even a *Götterdamerung* for science. In the short term they are giving rise to the formation of larger laboratories and institutes and a concentration on centralized facilities. In this chapter I argue that these developments are leading to dissipation of scientific activity with increasing managerial responsibilities, the curtailment of individual freedom in scientific decision-making and an increasing scientific bureaucracy. For a healthy new science there is a need to subject these trends to the democratic process and to prevent their use as instruments for interventionist politicians, defence establishments and private industries.

That instrumentation is crucial for the development of science was recognized by Bacon in the seventeenth century (see Bacon 1915 edition) who wrote that 'No star seems to have exerted greater power and influence in human affairs than the mechanical discoveries – printing, gunpowder and the magnet.' In the nineteenth century the approach was still optimistic and positive, assuming that every increase of power of apparatus is quickly followed by new discoveries and an increase in knowledge, or that 'Science owes more to the steam engine than the steam engine owes to science' (Conant, 1961). During the 1930s the increasing technology became critical for the new physics as reflected in the discovery of the neutron, the tremendous increase in power of proton synchrotrons and many other developments. Max Planck (1969) was one of the first to realize that this implied that 'with every advance in science the difficulty of the task was increased'. Scientific battles were to be won at ever-increasing costs. More recently we have seen that this also applies to what Nicholas Rescher (1978) calls complexity-intensive sciences such as quantum mechanics.

Similar developments have occurred in chemistry during the 1950s and 1960s with the introduction of nuclear magnetic resonance, electron spin resonance and other power-intensive spectroscopic techniques. In biological sciences the trends have been similar, but more recent. For example, expenditure on nuclear magnetic resonance, X-ray equipment and computer graphics consumed around 25 per cent of the biological sciences budget of the UK Science and Engineering Research Council (SERC) during the financial year 1983–84 (SERC Annual Report).

The overall position in the Science Board of the SERC is much more depressing when the central facilities are considered in addition to the University grants line of funds. For example, approximately £16 million is spent annually on neutron beams at the Institut Laue-Langevin in Grenoble and at the Neutron Spallation Source in the UK. An additional £8 million is committed to the Synchrotron Radiation Source and the Laser Facility. Together these total more than is spent on all SERC grants in mathematics, physics, chemistry and biology in all the universities in the UK. On top of this further sums of money are committed to other centrally co-ordinated initiatives, such as the development of single-user mini-computers. All of these developments are leading to an increasing percentage of the science budget being spent centrally, and a decreasing flexibility in the use of the funds available. The situation is further complicated by the very large sums of money spent by the Nuclear Physics and Astronomy Boards; these are dependent on governmental decisions and fluctuating exchange rates, and are to the advantage of a relatively small part of the science community.

The increasing cost of the equipment brings with it pressures to organize scientists into larger user-groups. The increasing complexity and interdisciplinary nature – especially in the chemical and biological sciences – also leads to more organizational structures. Both of these trends have important implications for the social organization of science.

First, there is an increasing trend of scientists associated with large equipment to become managers (Ravetz, 1971). Their managerial talents become more valued than their science, and support is often given in the name of scientists who are not experts in the particular technique, but have the ability to bring groups together. Secondly, the number of operations in a particular scientific area is limited. For instance, biological NMR, image processing and crystallography of molecular biology will probably be organized into not more than five groups in the UK, and synchrotron radiation is already organized in one centre in the UK with pressures to move to one group in Europe. Thirdly, these developments bring an increasing bureaucracy; the large establishments such as the Synchrotron Radiation Source may employ several hundred people, of whom only tens may be scientists who are outnumbered by adminstrators.

These trends have very serious implications for us as scientists. Rousseau has pointed out that the elite in any country tends to be proportional to the square root of the size of the community. Nicholas Rescher has suggested that the number of active scientific workers in an institute often varies in a similar way. Thus, the active scientist is increasingly in an atmosphere of bureaucracy and administrators – people who are not actually being scientifically creative – rather than in a stimulating research atmosphere. There are further problems in gaining access to the central facilities. These most often result from the limited number of persons who can use the machine at any time. However, the centralized control can lead to abuse of responsibility and consequent decrease of freedom for the individual scientist. Large institutions, their assembled structures and bureaucracies, can militate against certain individuals by denying them access to equipment. This is, of course, difficult to identify, but may happen more often than the average trusting scientist realizes.

The increasingly centralized organization of science can also be misused by government and industry. Even self-acclaimed non-interventionists such as President Reagan have been in dispute with their scientific communities, for example over the proposal of the administration for a new light source for materials scientists in the USA – a proposal that was politically attractive. Other government intervention may often have a more industrial or defence motive. For example, the enthusiasm in the UK to use Spacelab for scientific experiments on microgravity is clearly not based on the excellence of the proposed experiments. These have been very critically reviewed by peer groups. More probably there is commercial advantage for British companies in gaining space contracts in addition to the obvious longer-term defence implications. In fact the Conservative Government of Margaret Thatcher has often been interventionist in its science policy. For example, the Alvey Directorate in Information Technology has set up its programme by asking groups of industrialists to propose research objectives to which academics are enrolled. This pattern reappears in recent initiatives of the Biotechnology Directorate of the SERC. Such policies can distort academic research for often short-term industrial objectives or, even worse, quite unrealistic political aims. With such a policy recombinant DNA technology would never have been brought about.

In conclusion we face the future with an increasingly organized science with fewer opportunities for individual initiative. I believe that many of the trends towards centralization must be opposed strongly, but others are inevitably due to the fundamental nature of science itself. There can be no going back to small organizations in all areas of science. The only possibility is that we develop more open and conscious discussion of scientific issues, and that scientists participate more in the decisions that are increasingly taken *in camera*. The need has never been greater for a new social organization of science of the kind proposed by J. D. Bernal nearly half a century ago.

13

Towards a New Physics

John Taylor

I would like to try to explain why some scientists working in the science of physics are now conjecturing that we are seeing the end of fundamental science. The natural question to ask in response to this suggestion is: are we fantasizing or are we not? Scientists usually try to present ideas in such a way that they can be tested, and in this case I think it is true to say that we will ultimately be able to test these ideas, and very likely show whether they are right or wrong. At this point it is still only a conjecture that we are approaching a limit to science. We may know the true situation soon, and if it is the case that we are right, then clearly some very hard thinking has to be done by people working in fundamental science, in philosophy, in religion and in fact in any of the areas concerning knowledge.

To begin, let me take you back to 1864. In that year, James Clark Maxwell, from my own college, succeeded in unifying electricity and magnetism. His ideas were ahead of his time in the sense that they were not well accepted by numbers of his colleagues. Nor was it for some years that they were tested experimentally. We are now gaining the benefit of his unification in terms of the applications such a better under-standing of nature have given us: radio, television, radar, microwave ovens, optic fibres, etc. There are some dangers in these applications; we can get burnt up by laser beams. If you stand in front of a large microwave dish you can get heated up. Some of my colleagues working in microwaves even light their cigarettes by putting them in the beam!

Maxwell's unification of electricity and magnetism in 1864 produced the theory that physicists think is the most fundamental in physics in the last century. Where are we today as far as unification is concerned? The idea of unifying our understanding of the universe goes back to the ancient Greeks. We wish to explain the universe in as simple a way as possible. Maxwell's important step was continued in the mid-1970s when the second step was made. This culminated in the discovery at the big particle accelerator in Geneva, of what were called 'W and Z bosons', that some of you may have read about recently. That discovery proved that the next step of unifying the forces of nature had been successful, and that now electromagnetism and radioactivity had been unified in what is called the 'electro-weak' force. Before about 1975 there were four forces of nature: electromagnetism, radioactivity, the nuclear force and gravity. The unification of two of these had now occurred, leaving three forces: electro-weak, nuclear and gravitational.

The third step, that is unifying the electro-weak and nuclear force, has also been proposed. We are presently waiting for the latest results on the crucial experiment to test this unification. This critical test is whether or not the proton decays in about 10^{32} to 10^{33} years. That seems a very long time to have to wait, but by observing large amounts of material we can enhance the decay. If we see a single proton decay over a period of about a year in a large lump of material then we know that our further step of unification has been successful, and now three of the four pre-1975 forces will be unified.

We are now left with a final force, gravity, which is separate from these three other unified forces (usually called the grand unified forces). Gravity has not yet been included; but it is very different from the other forces. It is rather strange that primitive people would have known about gravity in its most immediate form when they fell off a cliff or fell out of a tree, or threw a stone into the air and noticed that it curved and fell back to the earth. But they knew little about electricity or magnetism, and nothing about the nuclear force or radioactivity. We know far less about gravity than the other forces of nature, and that is our problem at present.

Our problem about gravity is that Einstein, when he

introduced his idea that gravity was the curvature of space and time, did too good a job. He made gravity so elegant that it had to work that way – what more beautiful than that gravity was just curved space–time! Let me help you picture what that means. Imagine playing billiards on a plastic billiard-table top. If you place a large heavy black ball on the table, it pulls down the billiard table top, and you get curvature there that could be regarded as the gravitational force around the large ball. A small billiard ball then follows a curved path near the large ball, as if it were attracted to the big ball – that is gravity.

The ideas of Einstein are so simple that it is hard to see that they could be wrong. Indeed they have been carefully tested experimentally by various experiments in the solar system such as time-delays of radar signals from space vehicles, deflection of light from distant stars by the sun, etc. In all cases Einstein's theories have passed the tests with flying colours. We have to accept that gravity is identical to the curvature of space and time (to an accuracy of one part in a thousand, from the results of the above tests). But the other forces of nature are not curvature in space and time, because there is not anything left over for them to be the curvature of. We have curved space and time for gravity, so there is no space or time left to curve. If we want to unify the forces of nature, like gravity, the non-gravitational ones will also have to be curved. But these other forces that we seem to know so much about are not curvature, and so they are very different from gravity.

To avoid this impasse it has very recently been suggested that we may be living in a larger number of dimensions of space-time, not just the three of space and the one of time. There may be some extra dimensions which we have not yet been able to observe directly because effectively these dimensions are curved up into a very small ball. We have to get to very high energies before we can actually probe distances that are so small in this internal space. However these extra dimensions may well be manifested to us by the other forces of nature being curvature in this internal space. That is as attractive as Einstein's ideas on gravity. If it is true then we can indeed unify all the forces as curvature in this multi-dimensional space and time. We think now that 11 dimensions is the most suitable number for the world we live in, with seven extra dimensions beyond our known four. Seven is a magic number not only for physicists involved in these ideas, but also for Moslems who would believe in seven heavens.

It turns out that this approach can actually be used in any number of 'space–time' dimensions up to 11. If there are more dimensions than 11 then interacting particles must be included which are expected to introduce inconsistencies in their interactions (which latter are, however, necessary). Thus whilst seven is the magic number for extra dimensions there might only be any number between zero and seven. It is conjectured that if something is physically allowed then it should actually occur (termed the principle of maximality). That is why seven is the favoured number.

It is also possible that we only have a total of 10 dimensions, with six extra dimensions beyond our usual four. The 10-dimensional approach has become very attractive with a recent resurgence of interest in describing particles as extended objects like strings. Such a description can only be constructed satisfactorily in 10 dimensions since otherwise there would have to exist particles which have decidedly non-physical properties, and may even be 'ghost-like'!

When Einstein built his elegant theory of space, time and gravity, he suggested that gravity was to be considered as 'marble' – polished and beautiful – but that all the rest was to be regarded as 'wood', gnarled and twisted, since you can think of matter being any shape or size. But the problem that has always confronted physicists is why matter occurs in the actual guises that it does. We see it in a variety of shapes and sizes. Why has it had so many shapes, forms and different forces? We are suggesting now that we can see a way of pinning down the unique force that must occur between the various forms of matter in order that we can unify the matter, which is the wood, with the marble that is the gravity, to make it all marble in 11 dimensions. Indeed, that all is marble in 11 dimensions must be right – those of you who are sculptors will, I hope, agree. We would still have to understand more about what a 'unique' theory means and why we should suggest that our 11-dimensional theory could be so.

When we try to mix quantum mechanics (that we know governs the behaviour of ordinary matter), with gravity we find that the uniqueness of the 11-dimensional model is even more critical. It is possible, though has yet to be finally proved, that the only way that gravity can also be made to obey quantum mechanical laws, so giving a sensible theory of 'quantum gravity', is in the *unique* case of the curved 11-dimensional space–time. We seem to have a situation where, at least in the present framework of modern science, using the known forces of matter, the known forms of matter and gravity, and with quantum mechanics governing the forms of matter and gravity, that we are forced into an 11-dimensional theory. Everything is curved, as gravity tells us, and is uncertain, as quantum mechanics tell us. These two features, curvature and uncertainty, may only have a unique theory possessing them. We have not yet finally constructed that theory in the sense that we can make sensible predictions. That is the big goal that a number of groups around the world are trying to achieve. Whoever gets there first will have a very interesting result – the predictions that can then be made take us back to the beginning of the universe.

Let me now turn to consider historically what has happened in the development of fundamental physics. In particular let me consider how theories get discarded and replaced. It has always been that a theory has been presented, then tested, and then worked satisfactorily for a while until found wanting by more careful testing. New theories have come along, but at each stage the theory that was suggested had a whole variety of forms. Take Newtonian mechanics. There are many possible forces between the material particles, all obeying Newton's laws of motion. Thus Newtonian mechanics is ambiguous – it could not predict the forces. In any case Newton's laws could not be married satisfactorily with Maxwell's theory of electro-magnetism I mentioned at the beginning. This was only possible using Einstein's fundamental change in our conceptions of space and time from that of Newton, so that time was no longer absolute.

However, even these further developments associated with Maxwell and Einstein lead to a plethora of possible theories. None of these theories could be regarded as any sort of final or ultimate theory, since they could not answer the question: which of them (if any) is correct? There would always have to be some more precise theory which explains how to choose one of the range of theories as most suitable to describe the range of phenomena under consideration. We are now saying that we have a theory with no ambiguity. There is just one force law – nothing else. It would seem difficult to break that theory down using the methods I described above. That, indeed, is what I am now conjecturing – that we cannot conceive of a way of going beyond that unique theory. We have worked so hard to construct a sensible theory, and we find there is only one possibility. It would be remarkable if there were other frameworks beyond the present one we are living in, in which again there was only one theory. So some of us are thinking that indeed we have reached the ultimate theory of nature.

If one were a reductionist, for example (one who would understand problems of the behaviour of the whole in terms of the behaviour of its constituents), you then find that you have got to *the* fundamental aspect of nature in such a unique theory. Reductionism stops here. There are severe difficulties to raise at such a point. In particular how do we ever 'explain' the ultimate theory, if that is what it is? The only possible explanation, in terms of how science has progressed so far, would be for it not to be an ultimate theory. But if it is we must somehow see how a theory can be self-sustaining.

Let me turn from this to the questions we ask about the beginning of the universe. What about the nature of time itself? What happened before the Big Bang, a perennial problem which philosophers and theologians have asked again and again. It may well be that there are ways of seeing that time never really began in terms of a universe that has always been expanding. But I think we can begin to look with new eyes, in terms of a theory of this sort. Associated with modern developments in cosmology, our understanding of the way the universe began is rather different from the standard ways that have been put before us so far.

I think that what this whole discussion leads us to, is that fundamental physics will come to a halt, at least on the theoretical side. Experimentalists will have to work very hard, and of course this is very good for funding. As Tom Blundell says, problems in big science are becoming ever more important. Testing the ultimate theory, effectively from the beginning of the universe, will be *very* big science. The question of funding for that will be a little bit beyond the SERC. Certainly managerial features will begin to play an even greater role. For example, a point may be reached, after a long time with little further progress, that we accept that we have indeed reached the ultimate theory. There could then be some psychological disturbances among fundamental physicists because if they feel that the end has been reached, they will no longer continue with the quest.

I must emphasize here that I am not putting forward the proposition that there will be, or need be, no more creative mathematical thinking in physics. That such a proposal is false is clear from the high level of creative thinking going, for example, into problems in solid state physics. Moreover the theoretical/mathematical problems to be tackled in analysing any candidate for an 'ultimate theory' will be expected to be very difficult. There are already deep unsolved problems concerned with discovering the properties of the extension of Maxwell's electromagnetism to describe the properties of particles inside the nucleus. We should expect that a deeper theory than that would also have to solve these problems as well as further ones beyond them. The intellectual challenges to be overcome in solving the problems will be of the highest order. Their resolution might even lead to the glimpse of a theory even better than our ultimate theory! We might also say that in the end we are just at the beginning.

I may be charged with hubris at proposing that the end of fundamental science may be in sight. However theoretical physicists tilt not just at windmills but at the whole universe – sometimes they win! I can only finish by telling you an apocryphal story of Wolfgang Pauli, the famous theoretical physicist who suggested the existence of the particle we have never really detected – the neutrino – which has no mass, no charge, almost nothing. When he died he went to Heaven, as all good theoretical physicists do. When he was in the Presence he was asked: 'What would you like to know?' So he said, 'Well, how does it work?' 'Well', said the Presence, 'it would take a long time to explain it.' Pauli said, 'I've got all the time in the world', so the Presence started and he covered blackboard after blackboard after blackboard – finally on the 7th day he collapsed in a cloud of chalk and said: 'What do you think of it all?' Pauli said, 'Umm, very interesting – but weren't you wrong by a factor of 2 on the second line?' 'You're right' was the astonished reply. Naturally the response to Pauli's critical approach was not unimportant.

14

Feminism and Science

Janet Sayers

Women are participating today, perhaps more than ever before, in the movement for a new science that will meet the needs of all people – women and men alike. In the past the women's movement has contributed to this struggle primarily through asserting the ostensible old values of science against what was seen as their sexist abuse. It is now also increasingly challenging those values themselves, and arguing the need for them to be forged anew if science is to serve women equally with men. My concern here will be to describe this development in feminism's engagement with science.

The women's movement has, from its inception, taken issue with science in so far as it reflects and contributes to the maintenance of existing social inequalities between the sexes. At first this meant criticizing science solely in its own terms. This involved drawing attention to the way science departs from its own self-professed canons of neutrality in the sex discrimination it exercises against hiring and promoting women within its ranks. It also involved exposing the sexism that all too often governs the choice of research priorities, methods of investigation, and interpretation of results in science.

There are many examples of such departures from science's proclaimed social neutrality. As regards the sex discrimination exercised against women within the scientific profession there is, of course, the notorious case of Rosalind Franklin, whose contribution to the discovery of DNA was so singularly neglected by contrast with the attention accorded that of James Watson and Francis Crick.

This same sexism has also influenced the selection of research priorities within science. Thus, for example, the premise that women are naturally less capable of contributing to scientific research has led many investigators in the behavioural sciences to search for natural determinants of men's alleged superiority in scientific ability to the neglect of investigating sex similarities, or even the superiority of women as regards certain aspects of this ability. The search for a biological factor – a sex-linked gene, hormone, or brain difference, for instance – that might cause women's alleged natural inferior capacity for engaging in science continues today despite the repeated failure of any investigations to demonstrate that biology makes women less capable than men as regards those skills – visuospatial, analytical, or mathematical – associated with scientific ability (cf. Saraga and Griffiths, 1981; Sayers, 1982).

Likewise as regards research methodology. It has been shown that the presumption that men are naturally more aggressive and competitive than women has resulted in a focus on factors in male biology – androgens say – that might determine aggression to the neglect of equally studying factors in female biology – oestrogens say – that in fact also contribute to the determination of this behaviour (see Bleier, 1976). Overlooking the fact that such research thus breaches science's claim to be neutral and free from bias, the sociobiologist E. O. Wilson maintains that it provides scientific evidence of the justice of existing sexual inequalities within science's ranks. Citing this kind of research, he adds that science demonstrates that

> boys consistently show more mathematical and less verbal ability than girls on the average, and [that] they are more aggressive from the first hours of social play at age 2 to manhood (Wilson, 1975, p. 50).

On this basis he concludes

> Thus, even with identical education and equal access to all professions, men are likely to play a disproportionate role in political life, business and science (Wilson, 1975, p. 50).

Scientists have also been guilty of partiality in the interpretation of their data. Faced with the repeated finding that, on average, girls are verbally superior to boys as regards age of first speech and as regards verbal fluency, scientists – psychologists in this case – have interpreted these data as justifying the current unequal division of child-care between the sexes. Women's greater verbal ability, it is argued, renders them better fitted for child-rearing, for teaching their children to speak than men (see, e.g., Gray and Buffery, 1971). But this is to neglect the fact that such data also put in doubt the wisdom of the present sexual division of child-care in so far as it results in women's under-representation 'in political life, business and science' for which their greater verbal abilities would also seem to fit them better than men!

It continues to be important to challenge science on its own ground, and to draw attention to those instances in which it departs from its canons of neutrality in its professional practice, as well as in its selection and interpretation of evidence. For it is precisely the claim of scientists to neutrality that gives such weight and authority to the conclusions they draw from their data as regards the legitimacy of existing sexual inequalities in society. On the other hand, feminists have become increasingly aware of the need also to reassess the criteria by which scientific neutrality is assessed. It is to a brief account of this development within feminism that I shall now turn.

As indicated above, the traditional test of the neutrality of scientific theory is that it be free from contamination by social practice, in this case by that of sexual discrimination in society. But this very separation of theory from practice is itself implicated in women's social subordination, and in the failure of science to meet women's needs.

I shall seek to explain this point by way of an example – that of the history of the social management of childbirth. As many feminists have pointed out, this development involved the appropriation by medicine of many of the tasks formerly performed by midwives. Legitimation of this appropriation was often couched in terms that characterized midwifery as based on ignorant, unskilled superstition and customary practice. This it was said made it inferior to medical obstetrics based on scientific theory, and thereby held to involve superior technological skill and expertise (see, e.g., Oakley, 1976).

Clearly science has served women in freeing them from the abuses resulting from the hocus pocus of superstition as it informed midwifery and other areas of social life. On the other hand, the celebration of scientific theory over practical experience – the practical experience of women as midwives and as mothers say – has also operated against women's interests in so far as it has resulted in this experience being neglected in selecting the priorities and objectives of obstetric medicine. Consequently, as many feminists have pointed out, obstetrics – like other areas of science – fails to meet women's needs and interests as well as it might.

Feminists have accordingly often argued that the division of theory from practice, of mental from manual labour, must be transcended if science is equally to serve all people – women as well as men. In this, feminism is at one with socialism in its argument that, since this division constitutes one of the major determinants of the alienation and exploitation of the working class under capitalism, science will not serve the ruled equally with the rulers until this division is transcended in science as in other areas of social life (Rose, 1983). Socialists, however, all too often forget that if science is even-handedly to meet the needs of all people it must pay attention to women's needs just as much as to those of men. This, as feminists note, means not only transcending the division of theory and practice that currently characterizes science. It also means transcending the associated division of production from reproduction.

What is the bearing of production and reproduction, and their divisions, on feminism and science? In brief, and at the risk of over-simplification, I should explain that these divisions – both actual and ideological – can be shown to have been a major source of women's current social subordination. In the past, production was often conducted in conjunction with reproduction on a family basis within the home. The development of the forces of production – made possible in large part by the scientific advances of the late eighteenth and early nineteenth centuries – involved the development of machinery that was also initially operated on a family basis either inside or outside the home. Increasingly, however, it came to be operated outside the home on an individual basis. Productive activity thereby came to be separated from the reproductive activities involved in looking after the day-to-day physical and emotional needs of children and adults – activities that continued to be performed, as they are today, on a primarily family basis within the home.

While women of the working classes continued to be involved in both spheres of activity, middle-class women were discouraged from involvement in production. Increasingly they came to be identified with reproduction – with bearing and rearing the heirs to family property. Their menfolk, by contrast, were increasingly identified with production. This reflected the fact that it was mostly men who owned and administered the means of production and the wealth generated by it. Since it was in this sphere that wealth – at least exchange value – was generated, so production came to be valorized over reproduction. Furthermore, as a result of the ideological as well as economic dominance of middle- over working-class ideology, women of the working class, like women of the middle class, came to be identified with reproduction even though they were also involved in production; while men of all classes came to be identified with production (whether as workers or owners) and with the wealth it generates and the value thereby accorded it in society.

Since the division of production from reproduction has thus been a major source of women's work, needs, and interests being accorded less value, worth, and attention than those of men, so the women's movement has made the transcendence of this division one of its major aims. In the view of socialist feminism this entails seeking to integrate production with reproduction, not so much through domesticizing production as through socializing reproduction – namely child-care, education, health, and all those activities associated with the daily and generational reproduction of society's individual members.

What has all this to do with science? Scientific advance has been stimulated by, and has in turn stimulated, advance in the productive transformation of nature and its resources to meet human needs which have in their turn thereby been advanced and changed. Science has thus been very much allied with production and with the dynamic progress of society that contrasts with the resistance to change that is so often justified by an appeal to naturalism. As production came to be divorced from reproduction, and as production came to be identified with men, reproduction with women, so science too came to be increasingly regarded as masculine and its object – nature – as feminine in that the reproductive activity of women increasingly came to be equated solely with its natural, biological aspects (aspects that many scientists said would be endangered were women to engage in science). This view of science's object as that of nature equated with femininity is nicely illustrated by the historian of science, Ludmilla Jordanova, who describes in this context a 'statue in the Paris medical faculty of a young woman, her breasts bare, her head slightly bowed beneath the veil she is taking off, which bears the inscription "Nature unveils herself before Science" ' (Jordanova, 1980a, p. 54).

Although, as Jordanova (1980b) points out, science involves both subjective feeling as well as objective rationality, these aspects of its activity have tended to be polarized: the former aspect is deemed antithetical to, and therefore absent from, science and its pursuit of objectivity and neutrality. This polarization, I would argue, is itself an effect of the division of production from reproduction. For along with the ideological consequence of a division which equated men with production, women with reproduction, went a bifurcation of theory and practice, objectivity and subjectivity, reason and emotion, and of individuality and mutuality even though they are all involved alike in production and reproduction. As a consequence, production, and the science allied to it, came to be viewed as the seat of theory and of objective, individualistic rationality, and reproduction as the seat of practical concern, and of subjective emotion and mutuality. As feminists have pointed out, this ideological divide has in turn had baneful effects on science and its progress. The physicist Evelyn Fox Keller (1982) shows how it has resulted in many scientists – women as well as men – focusing on the individual operation of the objects they study to the neglect of their mutual interdependence and interaction.

The achievement of a science that does justice to the nature it studies, and of a science that serves women equally with men, depends on transcending antitheses such as those between individuality and interdependence that are in large measure a result and reflection of the ideological effects of the division of production and reproduction. It therefore depends on the integration of production with reproduction for which feminism strives. The struggle for a feminist science is thus one with the struggle of the women's movement in general. But it also poses peculiar problems of its own, of which I shall specify three below.

In the first place it depends on women gaining control over science's means of production in order that these means might be directed to meeting women's needs equally with those of men. This entails seeking to change the conditions that render women only skivvies and handmaidens to science – whether as factory hands making scientific equipment, domestic servants (whether at home or at work as cleaners, cooks, etc. to scientists), or as clerks, secretaries, technicians, and research assistants in scientific laboratories. It means wresting the control of scientific production from men so as to share equally with them in determining the priorities, objectives, and applications of science. Yet such is the association of science with men's interests that few women believe they have anything to gain from such struggle. This is the case despite the fact that science clearly shapes women's as well as men's activity in production and reproduction, in public as well as in private life (Haraway, 1985). Furthermore, even when girls and women recognize their interests to be bound up with science, and even when they seek to become scientists in their own right, they are often dissuaded from such participation on the ground that scientific and mathematical achievement depends on much more effort and hard work in the case of women than it does in the case of men (see, e.g., Parsons et al., 1982). This argument in turn reflects the prevalent view of science as men's natural preserve, one for which women are not naturally fitted and in which therefore their success depends on much more effort than it does for men.

A second obstacle to the struggle for a feminist science is that such a goal is often dismissed as a contradiction in terms.

The very essence of science, it is argued, lies in its neutrality. It should not therefore be allied to any political and social persuasion, feminist or otherwise. However, as indicated above, the development of science has been inextricably bound up with society and with the development and divisions of its productive and reproductive activities. Furthermore, despite its claims to the contrary, scientific research is also very much governed by subjective hunch and emotional inclination. Science can only fully make good its claim to objectivity – a claim that goes hand-in-hand with its claim to neutrality – if it acknowledges all the objective conditions of its own existence. This includes recognizing its political, social, subjective, and emotional determinants, and the lack of objective truth to the polarization of subjectivity from objectivity – a polarization that is more emphasized in the ideological elaboration of the division of production from reproduction than is warranted by reality. As the psychologist Jean Piaget once pointed out:

> Objectivity consists in so fully realizing the countless intrusions of the self in everyday thought and the countless illusions which result – illusions of sense, language, point of view, value, etc. – that the preliminary step to every judgement is the effort to exclude the intrusive self. . . . So long as thought has not become conscious of self, it is a prey to perpetual confusions between objective and subjective, between the real and the ostensible (cited by Keller, 1982, p. 594).

If science is to serve women it must recognize that this goal – that of meeting women's needs and interests as of society's in general – falls within its proper province and preserve. This means challenging the way in which its canons of objectivity and neutrality are used to mystify the actual relation of science to society (Fee, 1983). This is not to reject these canons in favour of a hopeless social relativism that would accord equal validity to sexism as to feminism in science. Quite the reverse! It is to argue the need to recognize the objective ways in which the sexism of our society affects its science both in its theory and in its practice, and it is to argue the need to acknowledge, in order to combat, the objective determinants of this sexism – namely the divisions both ideological and actual, of production and reproduction.

This brings me to a third, related obstacle to the development of a feminist science. Such is the force of the separation of production from reproduction, and of the related separation of science from society, that feminism is all too easily dismissed as peripheral to science, the main activity of which, as I have said, is conceptualized as essentially unrelated to social issues, feminist or otherwise. Moreover, even where it is acknowledged that science is related to society, it is more usually recognized to be related to production than it is to be related to reproduction. As a result, feminism and its concern with reproduction is still viewed as having little, if anything, to do with science.

Both the above ways in which feminism is marginalized with respect to science are thus related to the self-same divide of production from reproduction against which feminism strives. This marginalization is not due to any lack of numbers of women whose interests feminism claims to represent. It reflects women's lack of clout in a society where power, along with value, is accorded to social production and in which women, because of their role (actual or ascribed) in reproduction, enjoy relatively little status and authority.

For feminism to be given a proper hearing within science, let alone for it to achieve the transformation of science necessary to science's meeting women's needs equally with those of men, the division of production from reproduction must therefore be transcended. Until that happens feminism's engagement with science will, perforce, remain principally that of gadfly and irritant. As the biochemist and historian of science, Elizabeth Fee, puts it:

> At this historical moment, what we are developing is not a feminist science, but a feminist critique of existing science. It follows from what has been said about the relationship of science to society that we can expect a sexist society to develop a sexist science; equally, we can expect a feminist society to develop a feminist science. For us to imagine a feminist science in a feminist society is rather like asking a medieval peasant to imagine the theory of genetics or the production of a space capsule; our images are, at best, likely to be sketchy and insubstantial (Fee, 1983, p. 22).

This does not, however, entail that the struggle for a new, feminist science is utopian. The conditions for its existence are already in the process of being forged. Despite the divisions that exist between them, production and reproduction are indissolubly interlinked. Thus, for example, the hours of work men and women can devote to social production is conditioned by reproduction, by the provisions available, say, for the care of their children and other dependents while they work. Furthermore, production and reproduction are also in contradiction with each other. It is women's experience of this contradiction – their experience of the conflicting demands of work and home – that has constituted the impetus of the current women's movement. Consciousness of this contradiction has been heightened as an effect of women's increasing involvement in production – an involvement that is the result, in part, of scientific advance, of developments in contraceptive technology, for example, which have meant that women are not now removed from social production as they once were by the demands of constant childbearing. The conflicts between production and reproduction are thus even now operating to undermine and transform the divisions between them – divisions that I have been arguing constitute a primary source of the current failure of science and society to meet women's needs equally with those of men. In this lies cause for optimism. For it means that the conditions for the emergence of a new, feminist science – one that will equally serve women as well as men – are in the process of emergence not least because of the vigour with which the women's movement is now pursuing its cause.

15

Nothing Less than Half the Labs

Hilary Rose

It is obvious that the values of women differ from the values
which have been made by the other sex . . . yet it is the
masculine values which prevail.

Virginia Woolf.

The White House receives its advice from people who know
something about physics or chemistry. The person in charge of
biology is either a woman or unimportant. They had to put a
woman some place. They only had three or four opportunities,
so they got someone in here. It's lunacy.

James Watson quoted in *Science*, vol. 228, 12 April 1985.

The occupation of scientist or engineer is by and large secure,
technically interesting and reasonably well-paid; it is also still
very much a man's job. Consequently, there are good equal
opportunity reasons for wanting girls and women to have their
share of the scientific and technological pie. Equal-righters
have pressed energetically the claims of girls and women for a
fair share of both the educational provision and the jobs. Yet
for both radical feminists and socialist feminists, who carry a
more transformative conception of the need to change society
to make it serve the needs of women (and of men) better, the
equal rights position is seen as masking the real problem – the
oppressive nature of the scientific and technological pie.

Feminists have long been aware that science – and here I use
the word 'science' as a shorthand for 'science and technology' –

is not simply neutral but actively hostile to the interests of women; and that shorthand, science, for science and technology, is more than just a shorthand, for it recognizes, despite the cultural mystification which the separation of terms endeavours to impose, that science and technology are in any practical sense indivisible. Back in 1936, before the achievements of nuclear and solid state physics and the advent of the biological revolution had together made possible the bomb, biotechnology, the microchip and the test-tube baby, Virginia Woolf (1936) observed 'Science, it seems, is not sexless; she is a man, a father and infected too'. Those with a suspicion that Woolf was right in 1936 have even better reason to be concerned about the state of science and technology today. Do we really want our sisters to enter the juggernaut? Would success be but a Pyrrhic victory in which the sexual composition of the scientific labour force was changed but its character had not?

Here, I want to suggest that equal righters and transformative feminists[1] with their very different preoccupations about what feminists need to do, do not have to divide and oppose one another, and that in considering the question of women and science education, the reformist position of the equal-righters opens the possibility of a more radical programme of change. The issue is not 'are equal-righters or transformative feminists right?'. To set the question up this way is to enter the dichotomy beloved by masculinist reasoning – all that reform *or* revolution debate. The more useful question is whether the struggle to get women into science and technology is likely to help women and, if so, how? There is no invisible hand of feminism which guarantees a positive outcome, but posing the question this way helps us to look at the structural reasons for the exclusion of women from science which lie within the division of labour in a patriarchal society, and the consequences this exclusion has not merely for women but for the nature and goals of a masculinist science. How could we work for the entry of women into science and technology in such massive numbers that their sheer presence brings in those

---

[1] Despite theoretical disagreements between radical and social feminists both seek a radically different society – so I have called them 'transformative', as against the equal rights perspective.

different experiences, values, needs and demands which will themselves inform and transform the knowledge and applications of science?

Absent and invisible women

It is increasingly commonplace to talk about the absence, or at best limited presence, of women in science and engineering. The new chairman (*sic*) of the Equal Opportunities Commission, the marine engineer, Lady Beryl Platt, launched WISE (Women into Science and Engineering) as an initiative to encourage more girls to take science and engineering subjects at school so that the option of them making a career in these subjects is not precluded by choices made as far back as the 'O' level stage. In similar vein, WISE hopes to encourage industry to release its younger scientists and engineers, above all the women, to go to the schools to talk about the contribution women professionals can and do make. The EOC meanwhile provides posters of Hypatia's sisters, as it were, from the Alexandrian polymath Hypatia herself to Marie Curie and Dorothy Hodgkin, and glossy pamphlets showing smiling self-confident-looking young women wearing hard hats and standing on construction sites, dextrous and object-orientated in laboratories, or relaxed and totally in control of computing equipment. Positive role models in picture and actuality, together with innovatory teaching strategies, are seen as a way to alter the statistics of the gender distribution of scientists and engineers.

What does WISE have to take on in higher education? First, it is true, but only in a particular sense, that there are few women. The academic staff of science and engineering departments in Britain is almost entirely male, though the pattern varies from engineering, where there are almost no women staff and only 8 per cent women students, through physics and chemistry, which are a little but not much more mixed, to biology where (depending on the branch of biology concerned) there is the greatest proportion of women as academic staff, although still a small minority. What is special to biology, however, is that many departments have a majority of women undergraduates.

While there are proportionately few women in evidence in post-18 science and technology education, that is not to say that there are none to be seen at all. Women clean the floors – under the supervision of male supervisors; women act as technicians, under male senior technicians; they work as catering staff, under the direction of male catering officers; and they work as secretaries typing letters dictated by male academics and generally smoothing out interpersonal relations. The point is – and it has to be made again and again – that women's paid work, especially in the science or technology laboratory, echoes just what she does at home – except that at least there she is relatively free to get on with it at her own pace.

The very architecture itself reflects the routine expectation that this is a man's world. In engineering blocks built at the height of the 1960s affluence, when the norm was not a stretched and cut budget, it is terribly hard to find a women's loo. Whatever the Robbins principle in university expansion, it was not about equal opportunities for women. Even now, women are ambivalently wanted as students or colleagues. It is not just the pornographic pictures of women still pinned up in the labs. The new health and safety posters, for example, found in the labs and workshops are often crudely sexist in their use of women's naked bodies to make sure the men remember the safety points. The university which wrongly dismissed Mrs Dicks from her biology research post was at the same time discussing conversion courses with the EOC for women students who wished to do science or technology but had the wrong 'A' levels.[2] Even at the most superficial level, higher education is sending out contradictory messages to women about science and technology.

Scientific and technological education in Britain reflects the global, rather than merely national, class, race and gender order. Because the body of scientific knowledge is inherently international and the language of communication formal or even mathematical rather than verbal, the practitioners of science and technology need only the common culture of

---

[2] She was a biological research worker contesting her redundancy supported by the EOC against the same university which was discussing with EOC a conversion course to attract women into sciences.

industrialization. It comes as no surprise then that the engineering departments – even more than the science ones – recruit extensively among the rising elites of the newly industrializing societies. These Third World students are almost exclusively male; the much smaller numbers of women are to be found in the appropriately 'softer' and more feminine subjects. The indigenous British students are almost entirely male, but also almost entirely white. A handful of British Asians and almost no British Caribbeans thus completes the archetypal membership of the student body of a science or technology department at most British universities or poly-technics. Some institutions and departments have gone some way to contest this stereotype, and it is important to acknow-ledge their achievement; but they are a minority.

The point is that scientific and engineering education is indissolubly bound up with what science and technology are, and are used for, in our society. This is much more true for science and technology than it would be, say, for arts, history and many of the social sciences. Education under capitalism and patriarchy is fragmented because the division of know-ledge, division of labour and division of power are funda-mental to both. Capitalist patriarchy is not a monolithic universal category but one which has to be understood in terms of the variations between different societies over time. The deep antagonisms of class, gender and, for that matter, race find their reflection but with differing degrees of intensity within these societies. In certain respects, not least in the continuing resistance to the claims of women to have access to higher education, Britain is a particularly patriarchical society. (Under 40 per cent of higher education students are women, as against 50 per cent in most industrial countries.) It is difficult to escape the feeling that it is the relative shortfall of male school-leavers during the 1990s which has moved the male-dominated academy to view the possibility of increasing the numbers of women students in such a favourable light.

For that matter science and engineering are not monolithic either. Although there are some women academic staff in biology, and almost none in engineering, the exclusion mechanism works very differently between the disciplines. Biology appears to be the most sympathetic to women; even more girls than boys take biology; but by the time we are looking at the entry requirement into research – the PhD – the ratio of men to women is four to one. In engineering, physics and chemistry the exclusion mechanism seems to operate before higher education rather than during it. Thus any strategies for change have to be premised on the very clear understanding that the problems – while they have certain things in common – work out in different ways within the different disciplines.

The question of access into science and technology in Britain is exacerbated because of the early – indeed premature – specialization within the British school system, unlike, for example, the USA, where the commitment is to a broad general education until the end of the first degree. In Britain you either get a school science education, or you do not. Doing science at school is not about gaining knowledge of the culture of the society of which one is a member, but about taking the first step on the ladder to becoming a scientist or engineer. While the arts and social sciences have worked harder, and found it easier, to provide a second chance at study, universities in particular are remarkably resistant to thinking creatively and practically about a second chance of getting into the sciences. Anyone who has sat through a University Senate debate on the sacred value of three or four 'A' levels all in the natural sciences will know what I mean. Indeed, even with my hostility to Sir Keith Joseph's educational *dirigisme*, it is difficult not to have a sneaking sense of hope that the 'A' level stranglehold might be broken as at least one good thing among all the Joseph losses. Nor has the example of the Open University's success in teaching science and maths to students with no background done much to modify the ideological commitment to premature specialization. Such a commitment is bad news for all disadvantaged potential students; for women, because of the other barriers we now recognize which direct them away from the 'masculine' 'hard' subjects at school, it is extremely negative.

Science and technology under capitalist patriarchy

However, we cannot consider the question of increasing the participation of women in science without looking at what science and technology are actually doing in Britain today, and asking whether it is the kind of activity which women would want to enter.

The overwhelming majority of spending on science and technology by state and industry, and of the employment of scientists and technologists, can be categorized under two very traditionally male heads – science for military and internal security goals, and science for production and profit. Science for social welfare and for the environment come a very poor third (science for that disinterested goal – knowledge for its own sake – while important within the ideology of science, forms a very tiny part of the whole enterprise). You can see this in the distribution of the science and technology (research and development) budget, now some 2–2.5 per cent of GNP. The government's share of this, according to government sources, (HMSO 1983) was £3.38 billion in 1981–82 and about half was spent on military research and development. It is difficult to be very precise about these figures as a good deal depends on the detailed definitions employed; European Commission statistics, in 1983 for example, suggest that Britain spends 60 per cent of the state budget on military research and development.

What is discussed rather less than this is the fact that fundgivers, whether industry or the state, are, like the scientists and engineers who carry out the research, overwhelmingly male. Although varied in their class origins and positions they share certain common assumptions concerning the priority of the needs of the production of wealth and of military defence. It is by no means self-evident that productionism and militarism holds such a priority within the agenda of women. (It may, too, cast a little light on the fact that less than 10 per cent of the research on birth control technology is directed towards men (Leothars, 1984).

Historically, science as knowledge was seen as a liberatory force. Scientists were men (and rather infrequently women) critical of the social order, struggling for the greater freedom of the human spirit. A gendered look at that past might well interpret the freedom as being mainly gained for men and offering much less except in a rather abstract and relative way for women. However, even to speak of science as a critical force within culture is to invoke a period long since past. Today science is incorporated into the social order in a way that few other areas of the higher education syllabus are – with the possible exception of management and business studies (Rose and Rose 1976). All the evidence is that this incorporation is becoming ever more complete, with the passive support of the present generation of University Grants Committees and Research Council members and administrators.

Throughout most of their history, the development of physics, chemistry and engineering has been geared to the needs of the generation of weapons and profit. But the speed with which even the most arcane of scientific endeavours is pressed into service by state and industry in pursuit of these twin goals, although not fast enough for the technocratic imagination which believes that herein lies the core of the explanation for Britain's relatively poor economic performance over past decades, is still impressive. There is little space for autonomy in sciences ranging from astronomy, whose huge expenses are met with a view to the backhanders which may accrue to the space race and star war programmes, to biology, where what 20 years ago was an 'academic' study of genetics and the molecular components of life has become a scramble for patents and industrial investment as the great bio-technology bandwagon with all its promises (and threats) of genetic engineering (Warnock 1984) begins to roll.

If this is the character of contemporary science and technology, why should we want feminists (or indeed anyone) to enter it? But there is more yet. Go back to the golden age of critical science – or at least the myth thereof. It is doubtless true that the part science played in the transition from feudalism to capitalism was a vital one, and that this transition was part of a liberation of the human spirit and intellect from the dead weight of the received wisdom of church, king and custom. Exploring and understanding the natural world so as to generate the public knowledge which the methods of scientific enquiry yield were steps on the road to human freedom. But they were distinctly lopsided steps taken along a route which was simultaneously capitalist and male-dominated. Feudal society had, perforce, to endeavour to achieve some sort of harmonious relationship with a nature which it had not the means to control.

Critical social theory (Schmidt, 1973; Leiss, 1972), less enthusiatic about the role of science than the mainstream traditions of either reformist or Marxist socialism, argued that modern science had a fundamentally expoitative relationship to nature, seeking control and mastery. Domination replaced harmony. Through feminist scholars, notably Carolyn Merchant (1982), we have learnt to re-read Francis Bacon, acknowledging him as the major ideologue for both bourgeois and socialist cultural traditions in modern science, and see the sexual violence of his central metaphors. Bacon's science is lustfully masculine, his metaphors are unabashedly those of 'uncovering nature's nakedness . . . wresting her secrets from her . . . penetrating her inmost recesses'. They are the metaphors of rape. This explicitly male goal of domination and exploitation has remained central not merely to capitalist science as it has developed, but to the equally masculine traditional socialist view of the purposes, prospects and methods of science and technology.

We have had to wait for Evelyn Keller's (1983) study of the geneticist Barbara McClintock, who in 1984, at the age of 82, was awarded the Nobel prize for work done many years earlier (should we ask why so late as well as why so few?) in order to read a detailed account of a radically different relationship of a scientist to nature. McClintock speaks (and it is rightly reflected in the title of Keller's book) of *A Feeling for the Organism*. While McClintock made a salutory contrast to James Watson's (1984) macho account of the discovery of the double helix, Keller is not making an essentialist point concerning the differences between the perspectives. I could cite, for example, the geneticist J. B. S. Haldane who practised what he described as a non-violent biology. What feminism is after is a newly engendered perspective and *that* is not a biological phenomenon.

Science, technology and the new social movements

Bacon's exploitative view of science and nature has permeated the analysis of both Marxism and labourism through this century. Science and technology through the exploitation of nature were to do nothing less than end want. Such a conception of humanity's relationship to nature as one of mastery and exploitation lies at the heart of the labour movement itself. While not sharing the militaristic goals of a capitalist science and technology, the movement is deeply preoccupied with productionism through the exploitation of nature. It is precisely this conception which the new social movements of feminism and ecology challenge. Feminism claims women's rights over their own bodies, above all to choose to have or not to have children. Within feminism women look for a harmonious relationship between their lives and nature, beginning with their own nature. This harmonious relationship with nature intrinsic to feminism finds matching resonances with ecology. From an initial set of anxieties about 'limits to growth', finite natural resources, problems of global pollution and holocaust, an alternative vision of humanity and nature has developed.

To bring this new relationship into existence, both feminism and ecology necessarily reject the split, central to the science of domination, between the living participative 'I' of experience and the external world of nature. Where the old science cut thought and feeling in two, separated subjectivity from objectivity, the new movements seek a reconciliation (Harding and Hintikka, 1983). The objectivity of science which was claimed as its highest cultural achievement is seen by the new social movements as threatening human survival itself (Capra, 1983; Rose, 1982; Nowotny and Rose, 1979). It is uncomfortably like the old medical joke which similarly split technique from social purpose. 'Yes, the operation was a great success; unfortunately the patient died.'

The difficulty the new movements face in bringing this new conception into the old science and technology and transforming them is nowhere more evident than in the very different trajectories of ecology as a social movement and ecology as a subject within higher education. Born during the optimism of the 1960s and 1970s the teaching of ecology was seen as a creative response to the problems of pollution and the need to conserve natural resources. As a new strand within biology it found an uneasy place in school, further and higher education syllabi during the 1970s and attracted students – both men and women – who shared these goals.

Ecology as an academic subject is defined as the study of the relationships between communities of living organisms and the living and non-living world around them. Sustained by the growing public concern with ecological issues, the subject grew steadily. No small part of the credit for this new public concern must be attributed to Rachel Carson's path-breaking book *Silent Spring*. While not a self-identified feminist, her writing was to speak eloquently of the twin risks of human and natural catastrophe resulting from the present techno-logical culture. Today we have Bhopal.

But the sad thing is how far the educational courses have become separated from the utopian goals of the movement itself. The social movement stressed holism, the need to live non-violently and harmoniously with nature. The science and the movement look to emphasize the complexity and inter-connectedness of the living world; the belief that it is not possible to understand the present without its past; nor by disarticulating nature into molecular components alone; nor as divorced from human interest and intervention – and hence human values. But ecology as a taught science in the context of a deteriorating economy became trapped into precisely the fragmented, analytical mode of approach of the classical science disciplines. This captured ecology laid its stress on objectivity, on control, on reductionism, as opposed to the holistic understanding represented by the hopes of the social movement. Where the movement sought to transform the world, ecology courses increasingly turned out a generation of pollution control experts and planners to be snapped up by industry to provide the know-how to manipulate, circumvent and avert planning legislation and pollution control standards.

Feminism, the division of labour and a new vision of science

Why does this more harmonious vision of the relationship of

humanity and nature flow from feminist theory and understanding? Why is it so conspicuously lacking from actually
existing science and technology? Surely it is precisely the
division of labour between women and men in contemporary
industrial society which cuts women out of science (Rose,
1983). A division of labour which allocates caring and
reproductive labour to women and cognitive and productive
labour to men, is bound to result in the science done by the
men reflecting their male priorities and goals. Science can only
be transformed from an instrument of oppression and destruction into one of liberation for all humanity by its reorientation
towards feminist (which offers genuinely humanist) goals.
This is why the project of feminizing science is much more
than a simple EOC-type project however much we need to
encourage and struggle for equal opportunities strategies. To
be realist as well as visionary requires change on multiple
levels. Without such a multifaceted approach, projects like
WISE threaten to become new victims to what has been
unforgettably described in policy innovations as 'reinventing
the broken wheel'. Other broken wheels have followed the
post-Sputnik panic of the late fifties, when the USA tried very
hard to get more girls and women into science and technology
so as to compete more effectively in the superpowers space
war. In Britain 1969 was also an earlier year of getting women
into engineering. Puns apart, it might be wise to look at these
earlier initiatives in the light of the fact that there are
proportionately fewer women academics in, for example,
mathematics and physics now than there were in the 1920s. It
is only medicine as a science–based profession which has
made significant space for women. Indeed, in Britain it is
particularly critical to examine the relationship between
education and training and the labour market. Hakim's (1978)
work, for example, points to an increasingly segregated labour
market, with women taking a diminished share of the skilled
jobs, so that despite 60 years of educational reform, women's
position in the labour market was relatively worse in 1971 than
in 1911.

This critique of actually existing science and technology has
been root and branch. The need to transform science and
technology has never been more urgent, yet at the same time it

has never been more possible. The burgeoning social movements of ecology and feminism are challenging the existing order not merely in theory but in a multitude of local practical activitites. The socialist cities, London and Sheffield particularly, have fostered a mass of experimentation and innovation, releasing people's creative potential – take the work of the Greater London Enterprise Board (GLEB) with its programme of local computer and technically equipped workshops. East Leeds Women's Workshop, which has sought to give women technological skills while simultaneously offering child-care, has been an important pointer to the new directions that we need for a socialist policy in science. It is not surprising that this present government has set cash limits and has refused to support projects designed, like Leeds, to have integral child-care facilities.

Science and the demands on women's time

Women's lives are not organized around paid work like those of men; instead they are threaded between the demands of caring for children, running a home, caring for an elderly parent: each activity must be worked around the next. Extraordinarily few women can leave home in the morning, simply returning to relax at night to eat an evening meal and to sit by the box, read a book, perhaps go out for a drink, even to a meeting. It is this fundamental difference between most women's and men's lives which makes getting and *keeping* girls and women in science and technology so problematic.

Even studying science and technology in higher education regulates time in a less flexible way than other subjects, and as I have already said there is a deeper link between studying and doing science than there is in many of the disciplines. An arts or social science student can choose the most convenient place to study – library or home – the science or engineering student, like a factory worker has virtually to clock in. S/he has literally to be in the laboratory, to work through experiments, to analyse them on computer terminals. Much less can be taken home to be fitted around other activities.

What is true at the undergraduate level is even more true at the postgraduate level and post-doctoral level. At this stage the young scientist or engineer is expected to spend long hours in the laboratory; extended, even overnight, experiments are part and parcel of the process of becoming a recognized scientist. Even though it is for safety rather than a measure of work, science and engineering students have to sign in when they work unsocial hours. How does this world of exclusive preoccupation with science match up to the multiple demands a woman experiences? When in an account of the 'surplus graduates' Geoffrey Beattie (*Guardian*, 21 January 1984) writes of women biochemists being 'more realistic' and accepting jobs as lab technicians, what the author is really saying is that in order to limit the number of hours demanded of them by science, the subordinate job of a technician as seen as having the merit of defining and controlling the hours of work. Thus, the 'realism' of which Beattie speaks approvingly, is that women must in practice submit to the unjust and unequal division of labour inside and outside the laboratory.

Her double labour burden means that not only is she likely to find coping with the male pattern of scientific work difficult, but that the paucity of adequate nursery provision pushes her out altogether. There is not a single woman Vice-Chancellor or Principal in the entire British university system (nor for that matter a single woman Director of a Polytechnic), so it is not too surprising that Britain's Vice-Chancellors managed to conclude that universities did not have to provide nurseries and creches. It seems that it is all right to subsidize student accommodation, rugby, football, hockey, swimming pools, sports halls, drinking and catering facilities, but not creches. We are required to believe that the gender imbalance among the students using these facilities is simply a matter of chance. A much-needed piece of simple research is to measure the gender differential usage of student so-called universal facilities.

This lack of good facilities – from birth-to-5 creches, and play provision for 5+ children, affects all women students and staff with young children adversely, and those in the laboratory-based subjects, doubly so. Without child-care provision – a crucial element in the material conditions of their participation – they are denied access to academic and particularly scientific life.

Many women simply lack the possibility of carrying on in full-time science and engineering when they have children; others want to spend time their children. Part-time participation in science and engineering is extraordinarily difficult, and the tightening of the academic labour market has by and large eroded such limited opportunities as there were. If she is pushed or chooses full-time motherhood, getting back into science afterwards is tough. The knowledge base of science is cumulative; 5–7 years out and the re-entry into research is either impossible or at a very junior level.

There is an acute need for women to have adequate child-care facilities in higher education so as to make the care of children into a public and convivial activity, so that, for instance, a parent (or any other friendly adult for that matter) can nip across between lectures or experiments to play with children. This itself would help produce a different and more human environment for studying and researching. If we are serious about bringing women into science and engineering we have to bring the children in. But there is no joy in setting up nurseries so that women scientists become like men scientists. It is not simply that women rightly do not want to become men, but that if there is to be a chance of society getting the humane science and technology it so urgently needs, it can only be achieved by bringing into science those whose values are informed by their caring labour.

The provision of nurseries is thus both to make possible women's entry into science, but also, through encouraging men to share in child-care, to erode the overcommitment to the male scientific work ethic. The changing structure of employment is presently directed by a policy which deliberately requires the unemployment of 4 million people to achieve its economic and political objectives. Egalitarian objectives could seize the opportunities to fashion an entirely new relationship to the worlds of paid and unpaid work. The radical revision of the scientific work ethic which perpetuates the exclusivity of science is desperately overdue.

Reconceptualizing science education

The content and form of science and technology teaching need reconceptualization. Schools used to teach science as a set of received facts about the world, combined with so-called experiments – actually demonstrations designed to 'prove' or rather illustrate – these facts. (Even when the experiments failed, pupils were encouraged to write down what 'should' have happened.) Since the Nuffield science and education schemes some schools have adopted versions of its explicitly Popperian hypothesis falsification model of science teaching – though there are some indications that it disadvantages working-class children. Many teachers are still uneasy about this divorce of science from lived experience and social context, but are trapped by the rigidities of premature specialization, 'A' level and university entry requirements. Socialist and feminist science teachers could and should be encouraged to build new syllabi based around pupils' 'lived' experience, and to break the barrier between 'science' and 'non-science'. The now clear evidence that girls do better in science subjects and maths if taught in single sex classes should be recognized (Deem, 1978; Shaw, 1976, 1980).

Higher and further education build on the schools' approach. They receive 18-year-olds accustomed to learning from authority, to separating fact from value, and to the abstract world of taught science. The rigidity of the entry requirements to these subjects make polytechnics and universities inflexible to the needs of mature students, likely disproportionately to be women. Within the hallowed walls the authority of the research and review paper substitute for that of the teacher. The hierarchies of theory and practice, (hard) and mathematical sciences over (soft) and human-centred ones are reinforced.[3] Education is pragmatically directed towards, for the brightest, the research career leading linearly to fellowships of the Royal Society and the Nobel prize; for the rest, into an industry whose unidimensional goals are those of profitability which the student is not encouraged – or given the intellectual tools – to evaluate. Indeed, this cutting adrift of science from social context has now been specifically reinforced by Sir Keith Joseph who has frowned upon those few

---

[3] The sexual connotation of 'hard' and 'soft' remains acknowledged only at the level of jokes by 'malestream' philosophy of science.

'science and society' courses which were established – mainly in the polytechnics – during the 1970s.

To transform the content of this science and technology, in a way which makes it accessible to women and open to feminist and socialist demands, must also begin by opening the syllabus towards a more human and social-centred approach. This means more than tacking a few obligatory culture lectures on to the slack parts of the week along with the traditional Wednesday afternoon off for games. We must encourage science to be taught in its context – linking genetics with the implications of genetic engineering and birth control technology, biochemistry with the drug industry and biotechnology. We need to be aware of the pressures on students to avoid discussing the perspectives of science and technology in a social context; even where courses are available it is far from unknown for them to vote with their feet.

Getting large numbers of women into science and technology is to bring in that gender perspective stemming from the experience of those whose lives are interwoven between the demands and skills of caring and the demands and skills of scientific research and teaching. It may well be the only practical way of realizing the objective of changing science education from narrow technicism to a socially contextualized approach.

We need to link formal education to the rich ferment of community and trade union-based experimentation and innovation. The women's health movement is a significant force not only for changing health care but also for the accelerated self-education of women in the biology of women. Trade union groups seeking, like the Lucas workers, for socially responsible products as a means to save jobs and produce something worthwhile, similarly serve to release the creativity and self-learning of groups excluded from traditional ways of planning science and technology. The hermetically sealed world of science and technology education and research, opened or closed by the selection of subjects at 16, could be perfused with such new ideas. Of course there would be resistance, but the remarkable way which the sticks and carrots offered by a New Right government have forced education to the right suggest that another equally determined government could, and with rather more popular support from within education, move it towards feminist objectives.

Nothing less than half the labs

It is not possible in any discussion of getting women into science and engineering – that is, to be satisfied with nothing less than half the labs – to think that this can be achieved without radical changes at a multiplicity of levels. It is the overwhelming strength of great social movements that they do and can operate in this way.

Opening science and technology to serve new and more complex goals is no small task, and it is going to produce anxiety and resistance among educators who may well feel with some reason that they have something to lose. Restructuring from the left cannot be carried out with the same indifference to the human costs as the ongoing restructuring carried out by the right. But we have to acknowledge that much of what presently passes for science and technology may prove unnecessary or unwanted, and the training of scientists and engineers for these tasks quite inappropriate. Do we want aeronautical engineers or pharmacologists, for instance, if our goals are those of – say – the development of methods and designs to convert a large part of the existing housing stock – obsolescent to the ways that women, men and children live now – into flexible homes that can accommodate new relations of domestic life?

To sum up, the present state of science and technology and its education is that of training predominantly men to strive for the unidimensional goals of creating weaponry and profit. For much of the left it is enough to encourage men and women to enter a science and technology designed to promote economic growth. While this is a welcome improvement on the present commitment to militaristic science, we must go beyond this to begin to respond to the demands both of the old labour movement and of the new social movements confronting the social and economic crisis of our time. We need science and technology for economic sufficiency *and social growth*, a science and technology for women and men which will enable us to construct tools for conviviality and for the control and direction of our own lives. *Nothing less than half the labs* makes a good starting slogan.

Notes and References

Chapter 4

Maynard Smith, J. (1975) *The Theory of Evolution*, Penguin.
—— (1985) *The Problems of Biology*, Oxford University Press.
Monod, J. (1972) *Chance and Necessity*, Collins.
Ridley, M. (1985) *The Problems of Evolution*, Oxford University Press.

Chapter 5

Bateson, W. (1894) *Materials for the Study of Variation*, Cambridge University Press.
de Beer, Sir Gavin (1971) *Homology: An Unsolved Problem*, Oxford University Press.
Dobzhansky, Th. (1973), 'Nothing in biology makes sense except in the light of evolution', *American Biology Teacher*, March 1973, 125–9.
Goodwin, B. C. (1984) 'A relational or field theory of reproduction and its evolutionary implications'. In *Beyond Neo-Darwinism*, M.-W. Ho and P. T. Saunders (eds), Academic Press.
—— and Trainor, L. E. H. (1985) 'Tip and whorl morphogenesis in *Acetabularia* by calcium-induced strain fields', *Journal of Theoretical Biology* (at press).
Morata, G. and Kerridge, S. (1982) 'The role of position in determining homoeotic gene function in *Drosophila*', *Nature*, **300**, 191–2.
Odell, G., Oster, G. F., Burnside, B. and Alberch, P. (1981) 'The mechanical basis of morphogenesis', *Developmental Biology*, **85**, 446–62.
Oosawa, F., Kasai, M., Hatano, S. and Asakura, S. (1966) In *Principles*

of *Biomolecular Organisation*, G. E. W. Wolstenholme and M. O'Connor (eds), 273–303, Little, Brown & Co.

Oster, G. F., Murray, J. D. and Harris, A. (1983) 'Mechanical aspects of mesenchymal morphogenesis', *Journal of Embryology and Experimental Morphology*, **78**, 83–125.

Roberts, J. M. and Verrell P. A. (1984) 'Physical abnormalities in the limbs of smooth newts (*Triturus vulgaris*)', *British Journal of Herpetology*, (forthcoming)

Williams, N. E. (1984) 'An apparent disjunction between the evolution of form and substance in the genus *Tetrahymena*', *Evolution* **38**(1), 25–33.

Chapter 6

Bateson, P. (1983) 'Rules for changing the rules'. In *Evolution from Molecules to Men*, D. S. Bendall (ed), Cambridge University Press.

Dawkins, R. (1976) 'Hierarchical organisation: a candidate principle for ethology'. In *Growing Points in Ethology*, P. P. G. Bateson and R. A. Hinde (eds), Cambridge University Press.

—— (1982) *The Extended Phenotype*, W. H. Freeman/Oxford University Press (paperback).

Gould, S. J. (1978) *Ever Since Darwin*, Burnett.

Lewontin, R. C. (1979) 'Sociobiology as an adaptationist program', *Behavioral Science*, **24**, 5–14.

Medawar, P. B. and Medawar, J. S. (1984) *Aristotle to Zoos: a philosophical dictionary of biology*, Weidenfeld & Nicolson.

Rose, S., Lewontin, R. C. and Kamin, L. J. (1984) *Not in our Genes*, Penguin.

Teilhard de Chardin, P. (1959) *The Phenomenon of Man*, B. Wall (trans), Collins.

Tinbergen, N. (1953) *Social Behaviour in Animals*, Methuen.

—— (1968) 'On war and peace in animals and man', *Science*, **160**, 1411–18.

Wilson, E. O. (1975) *Sociobiology: the new synthesis*, Harvard University Press.

Chapter 7

Alexander R. D. (1980) *Darwinism and Human Affairs*, Pitman.

—— and Borgia, G. (1978) 'Group selection, altruism, and the levels of organization of life', *Annual Review of Ecology and Systematics*, **9**, 449–74.

Bateson, P. P. G. (1976) 'Specificity and the origins of behaviour', *Advances in the Study of Behaviour*, **6**, 1–20.

—— (1982) 'Behavioural development and evolutionary proccesses'. In *Current Problems in Sociobiology*, King's College Sociobiology Group (eds), Cambridge University Press, 133–51.

—— 1983a) 'Optimal outbreeding'. In *Mate Choice*, P. Bateson (ed.), Cambridge University Press, 257–77.

—— (1983b) 'Rules for changing the rules'. In *Evolution from Molecules to Men*, D. S. Bendall (ed.), Cambridge University Press, 483–507.

—— (1984) 'The biology of cooperation', *New Society*, **68**, 343–5.

Charlesworth, B. (1978) 'Some models of the evolution of altruistic behaviour between siblings', *Journal of Theoretical Biology*, **72**, 297–319.

Darwin, C. (1859) *On the Origin of Species*, Macmillian.

Dawkins, R. (1968) 'The ontogeny of a pecking preference in domestic chicks', *Zeitschrift für Tierpsychologie*, **25**, 170–86.

—— (1976) *The Selfish Gene*, Oxford University Press.

—— (1981) 'In defence of selfish genes', *Philosophy*, **56**, 556–73.

—— (1982) *The Extended Phenotype*, Freeman.

Eibl-Eibesfeldt, I. (1970) *Ethology: The Biology of Behavior*, Holt.

Hinde, R. A. (1984) 'Why do the sexes behave differently in close relationships?', *Journal of Social and Personal Relationships*, **1**, 471–501.

Humphrey, N. and Lifton, R. J. (1984) *In A Dark Time*. Faber & Faber.

Kropotkin, P. (1902) *Mutual Aid: A Factor of Evolution*, Heinemann.

Lorenz, K. (1965) *Evolution and Modification of Behavior*, University of Chicago Press.

Lumsden, C. J. and Wilson, E. O. (1981) *Genes, Mind, and Culture*. Harvard University Press.

Maynard Smith, J. (1982) 'Introduction'. In *Current Problems in Sociobiology*, King's College Sociobiology Group (eds) Cambridge University Press, 1–3.

Oyama, S. (1985) *The Ontogeny of Information: developmental systems and evolution*, Cambridge University Press.

Partridge, L. (1983) 'Genetics and behaviour'. In *Animal Behaviour*. Vol. 3. *Genes: developmental and learning*, In T. R. Halliday and P. J. B. Slater (eds), Basil Blackwell, 11–51.

Rose, S., Lewontin, R. C. and Kamin, L. J. (1984) *Not in our Genes: biology, ideology and human nature*, Penguin.

Thorpe, W. H. (1961) *Bird-Song*, Cambridge University Press.

Rowell, C. H. F. (1971) 'The variable coloration of Acridoid grasshoppers', *Advances in Insect Physiology*, **8**, 145–98.

van den Berghe, P. L. (1983) 'Human inbreeding avoidance: culture in nature', *Behavioural and Brain Sciences*, **6**, 91–123.
Westermarck, E. (1891) *The History of Human Marriage*, Macmillan.
Wilson, E. O. (1975) *Sociobiology: The New Synthesis*, Harvard University Press.
—— (1976) 'Author's reply to multiple review of "Sociobiology" ', *Animal Behaviour*, **24**, 716–18.
Wrangham, R. W. (1982) 'Mutualism, kinship and social evolution'. In *Current Problems in Sociobiology*, King's College Sociobiology Group (eds) Cambridge University Press, 269–89.

Chapter 8

Boden, M. A. (1977) *Artificial Intelligence and Natural Man*, Basic Books.
Mayhew, J. and Frisby, J. (1984) 'Computer vision'. In *Artificial Intelligence: Tools, Techniques, Applications*, T. O'Shea and M. Eisenstadt (eds) Open University, chap. 10.

Chapter 11

Doll, Richard and Peto, Richard (1981) *The Causes of Cancer*, Oxford University Press.
McKeown, Thomas (1976) *The Role of Medicine*, The Nuffield Provincial Hospital Trust.
—— (1979) 'The direction of medical research', *Lancet*, **ii** (8155), 1281–84.
Virchow, R. Undated reference to Einheits-Bestrebunger in *der Wissenschaftlichen Medizin*, cited in *Familiar Medical Quotations*, Maurice B. Strauss (ed), Little Brown & Co.

Chapter 12

Bacon, F. *Novum Organum*, Bk I, Sect. ii. See edition by Gough, A. B., Oxford, Clarendon Press 1915.
Bernal, J. D. 'Social functions of science'.
Conant, J. B. (1961) *Science and Common Sense*.
Planck, M. *Vorträge und Erinnerungen*.
Ravetz, J. R. (1971) *Scientific Knowledge and its social problems*, Oxford, Clarendon Press.
Rescher, N. (1978) *Scientific Progress*, Basil Blackwell.

Chapter 14

Bleier, R. (1976) 'Myths of the biological inferiority of women: an exploration of the sociology of biological research', *University of Michigan Papers in Women's Studies*, **2**, 39–63.

Fee, E. (1983) 'Women's nature and scientific objectivity'. In *Woman's Nature: Rationalizations of Inequality*, M. Lowe and R. Hubbard (eds), Pergamon Press.

Gray, J. A. and Buffery, A. W. H. (1971) 'Sex differences in emotional and cognitive behaviour in mammals including man: adaptive and neural bases', *Acta Psychologica*, **35**, 89–111.

Haraway, D. (1985) 'A manifest for cyborgs: science, technology and socialist feminism in the 1980s' *Socialist Review*, *15*(2), 65–107.

Jordanova, L. (1980a) 'Natural facts: a historical perspective on science and sexuality', In *Nature, Culture and Gender*, C. P. MacCormack and M. Strathern (eds), Cambridge University Press.

—— (1980b) 'Romantic science? Michelet, morals, and nature', *British Journal for the History of Science*, **13**, 44–50.

Keller, E. (1982) 'Feminism and science', *Signs*, **7**(3), 589–602.

Oakley, A. (1976) 'Wisewoman and medicine man: changes in the management of childbirth'. In *The Rights and Wrongs of Women*, A. Oakley and J. Mitchell (eds), Penguin.

Parsons, J. E., Adler, T. F. and Kaczala, C. M.(1982) 'Socialization of achievement attitudes and beliefs', *Child Development*, **53**(2), 310–39.

Rose, H. (1983) 'Hand, brain and heart: a feminist epistemology for the natural sciences', *Signs*, **9**(1), 73–90.

Saraga, E. and Griffiths, D. (1981) 'Biological inevitabilities or political choices? The future for girls in science'. In *The Missing Half*, A. Kelly (ed.), Manchester University Press.

Sayers, J. (1982) *Biological Politics: Feminist and Anti-Feminist Perspectives*, Tavistock.

Wilson, E. O. (1975) 'Human decency is animal', *New York Times Magazine*, 12 October, 38–40, 42–6, 48, 50.

Chapter 15

Capra, Fritijhof (1983) *The Turning Point: Science, Society and the Rising Culture*, Bantam.

Deem, Rosemary (1978) *Women and Schooling*, Routledge and Kegan Paul.

—— (1980), (ed), *Schooling for Women's Work*, Routledge and Kegan Paul.

Hakim, C. (1978) *Occupation or Segregation*, Research Paper no 9, DOE, November, pp. 1264–68.

Harding, S. and Hintikka, M. B. (1983) (eds), *Discovering Reality*, Reidel.

HMSO (1983) *Annual Review of Government-Funded R & D*.

HMSO (1984) *Report of the Committee of Inquiry into Human Fertilisation and Embryology* (Warnock Report), Cmnd 9314.

Keller, E. (1983) *A Feeling for the Organism*, Freeman & Co.

Kelly, A. (1978) *Girls and Science: An International Study of Sex Differences in School Science Achievements*, Almquist & Wiksell.

—— (1981) (ed) *The Missing Half*, Manchester University Press.

Leiss, W. (1972) *The Domination of Nature*, George Barziller.

Leothars, A. (1984) 'Inequalities in health care and birth control provision in the UK', paper given at the Social Administration Association Conference.

Merchant, C. (1982) *The Death of Nature*, Wildwood House.

Nowotny, Helga and Rose, H. (1979) (eds) *Counter-movements in the Sciences*, Reidel.

Rose, H. (1983) 'Hand, brain and heart: towards a feminist epistemology for the natural sciences'. In *Signs: International Journal of Women in Culture and Society* 9(11), Fall, pp. 73–90.

—— and Rose, S. (1976) 'The incorporation of science'. In H. Rose and S. Rose (eds) *The Political Economy of Science*, Macmillan.

Rose, S. (1982) (ed) *Against Biological Determinism*, Allison & Busby.

Schmidt, A. (1973) *The Concept of Nature in Marx*, New Left Books.

Shaw, Jenny (1976) 'Finishing school: some implications of sex-segregated education'. In D. Leonard Barker and S. Allen (eds) *Sexual Divisions and Society*, Tavistock.

—— (1980) 'Education and the individual: schooling for girls or mixed schooling – a mixed blessing'. In R. Deem (ed) *Women and Schooling* (1978).

Walden, R. and Walkerdine, V. (1981) *Girls and Maths: The Early Years*, Bedford Way Papers 8, University of London, Institute of Education.

Warnock Report (1984) *see* HMSO.

Watson, J. (1968) *The Double Helix*, Weidenfeld and Nicolson.

Woolf, Virginia (1936) *Three Guineas*; reprinted, Penguin.

Index

absent women, 181–4
accelerators, particle, 28, 162
access·
  to health care, 149–8
  to scientific facilities, 159
accidents, 146, 152–3
  genetic, 151
adaptation, 40, 43, 82, 83–6, 91
adaptive response of doctors, 149
affluence, diseases of, 145
aggression, 170
AIDS, 22–3
Alexander, R.D., 84, 94
alleles, 80, 81
altruism, 66, 81
analagous form, 48–9
animals, 63–9, 91
  co-operation, 84–5, 87
  experiments on, 34–5
  vertebrates, 42, 43–5, 50
ante-natal diagnosis, 25
arms race, 96–8
artefactual intelligence, 13,
  115–30
  alternative strategies, 121–2
  computer revolution, 120–1
  history of, 118–19
  improvement of artificial
    intelligence, 127–30
  intelligence, 119–20

machines and brains, analogy
  between, 115–18
neuroscientists and brain,
  125–7
seeing, 124–5
selective attention, 122–4
artificial intelligence
  and artificial brains, 12–13,
    103–14
  improvement of, 127–30
artificial selection, 82
attention, selective, 122–4

Bacon, F., 26, 157, 186–7
Bateson, P., vii, 70, 71, 76, 78
  on sociobiology, 12, 79–99
Bateson, W., 50
Beattie, G., 191
behaviour
  evolution of, 11–12
  and genes, 88–9
  social, 64–8
Berg, P., 21
Bernal, J.D., 2, 27, 160
big science, 158–60, 166–7
biology, 5–6
  ecology, 187–9
  as historical science, 10, 47–60
    genes and morphology, 51–3
    history versus logic, 47–51

natural selection and
emergence of novelty,
59–60
organisms and fields, 56–9
reproductive process, 53–6
limitless vista, 8, 19–25
machines for, 159
structuralism versus selection,
9, 39–46
women in, 183
birth control, research for, 185
Bleier, R., 170
Blundell, T., vii, 166
on new science, 14, 157–60
Boden, M., vii
Gregory on, 131–3
on brain, 13, 103–14
Wall and Safran on, 115–30
Borzia, G., 84
brain
artificial, 12–13, 103–14
and neuroscientists, 125–7
and machines, analogy
between, 115–18
Buffery, A.W.H., 171
bureaucracy, scientific, 157,
159–60
Bush, V., 27
Butler, S., 27

cake metaphor, 88–9
cancer
mortality, 144–5
research, 21–3, 30
capitalist patriarchy, science
under, 183, 184–7
Carson, R., 188
centralization of science, 158–60
Charlesworth, B., 81
chemistry, machines for, 158
child
birth, management of, 171–2
care, 192–3
class
production and reproduction,
173
science education and, 183

classification, 42
problems, 62–4
clinical science, 148
coadaptation of genes, 83–6
coding for altruism, 81
competition, 12
politics and, 86–8
war and, 96–8
complexity, increasing, 157–60
computers, 12–13
characteristics of, 104–5
costs of, 158
history of, 118–19
improvement of, 127–30
revolution, 120–1
selective attention, 122–3
vision of, 106–14, 115, 119,
121–2, 124–5
see also brain
Conant, J.B., 158
conflict, social, 87–8
conformity, social, 94
conservation in evolution, 44
co-operation
politics and, 86–8
symbiotic, 84–5
costs
of health care, 136–7
of science, 27–9, 157–60, 184–5
Craik, (scientist), 126
Crick, F., 8, 24–5, 170
critical science, 185–6
critical social theory, 186–7
culture, 3

danger of science, 20
Darlington, C.D., 76
Darwin, C. and Darwinism, 4, 9,
23, 47, 50, 82–3
on facts, 4
and free market theories, 86
on homology, 50
neo-, 9, 10
sociobiology and, 65–78
see also evolutionary theory;
natural selection
Dawkins, R., vii, 31, 76

Bateson on, 80–4, 90, 92
  on sociobiology, 11, 12, 61–78
de Beer, Sir G., 48, 51
de Solla Price, D., 28
death *see* mortality
defeatist response of doctors, 149
Descartes, R., 118
destabilization of society, 23–4
determinism, 11
  Dawkins on, 76–8
  genetic, 76–7, 79–80, 88–92, 93
developmental arguments, 91–2
diagnosis, development of, 153–4
Dicks, Mrs, 182
differences, transmission of
    inherited, 54
dimensions of space-time, 163–5
diseases, 22–3, 148
  see also infection
division of labour *see* production
    and reproduction
DNA, 8, 20, 21, 31, 168
  changes in, 40, 44
  recombinant, 20, 21–2, 25
  self-replication of, 54
  survival of, 81
Dobzhansky, T., 197
doctors, 148–508
Doll, R., 145
Drosophila, 40–1, 51–2, 59
drugs, therapeutic, 146, 147,
    154
duration of life, 151

ecology, 187–9, 190
education, health, 143, 152
education, science
  specialization in, 184, 190
  for women, 181–4, 194–6
    labour market and, 190
    reconceptualizing, 193–5
Eibl-Eibesfeldt, I., 89
Einstein, A., 97, 163–5
elderly, health of, 145, 146, 148,
    156
electricity and magnetism,
    united, 161–2, 164, 167

united with radioactivity, 162
electro-weak force, 162
elements, transmutation of, 49
emergence of novelty, 59–60
employment
  accidents and morbidity, 146,
    153
  in science for women, 179
  see also women
engineering, women in, 181,
    190
environment
  adaptation to, 91
  and health and mortality, 145,
    151–2, 155–6
  selection and, 9
EOC *see* Equal Opportunities
    Commission
epidemiology, 148
Equal Opportunities
    Commission, 181, 182, 189
equilibrium, punctuated, 10
ethical issues in science, 33–5
euthanasia, 24–5
evocative analogy, 116–17
evolution and social
    prescription, 92–6
evolutionary theory, 9–11
  see also, biology, as historical
    science; Darwin; natural
    selection

facts and hypotheses, 4
Fee, E., 176, 177
feminism *see under* women
fertilization, *in vitro*, 33–4
fields and organisms, 56–9
Fieser, L., 29
Fifth Generation Project, 13
Fisher, Sir R., 9
food, supply and health, 142–3,
    156
forces of nature, unification of,
    161–6
form *see* morphology
Franklin, R., 8, 169
free market theories, 86, 87

future
  health, 151–3
  of science, 19

Galen, 118
Galvani, L., 117
gender *see* sex
genes/genetic
  accidents, 151
  'blueprint', 89
  and behaviour, 88–9
  coadaptation of, 83–6
  determinism, 76–7, 79–80,
    88–93
  engineering, 20, 21–2
  and morphology, 51–3
  and selection, 10, 11
  selfish, 5, 80–2
  and social organization, 5
genotypes, 51, 84
Goodwin, B., vii, 55
  on biology, 10, 47–60
Goudschmidt, R., 44
Gould, S.J., 66
government misuse of science
  facilities, 160
gravity, 162–5
Gray, J.A., 171
Gregory, R., vii, 116
  on minds, machines and
    meaning, 13, 131–3
Griffiths, D., 170

Habermas, J., 3
Hakim, C., 190
Haldane, J.B.S., 9, 187
Haraway, D., 175
Harvey, W., 117
health, 14, 137–56
  education, 143, 152
  implications, present and
    future, 151–3
  medicine and
    capability of, 153–5
    childbirth, management of,
      171–2
    impact of, 138–40
  practice of, 147–51
  women in, 190
  mortality trends, 139, 140–1
    explanation of, 141–6
  sickness, non-fatal, trends in,
    146–7
heart disease, 144–5
Heisenberg, W.K., 118
Hinde, R.A., 93
historical science *see under*
  biology
history
  of artefactual intelligence,
    119–20
  of health care, 138–9
  logic, versus, 47–51
holism, 29, 188–9
home, women's work in, 191–2
homoeotic, mutation, 51, 52
homology, 48–53
  see also morphology
'hopeful monsters', 44
human politics *see under*
  sociobiology
Humphrey, N., 97
Huxley, T.H., 41
hypotheses and facts, 4

ideological limits to science, 27,
  30–6
immunisation *see* vaccination
incest avoidance, 70–2, 93–5
individual freedom of scientists,
  curtailed, 157, 159–60
individualism, 87, 88
inequalities *see* women
inevitability of genetic
  determinism, 77
infection
  improved treatment for, 145–6,
    154
  mortality from, 141–2, 143
inferiority of women, 170, 172
information technology *see*
  computers
inheritance, 47, 54, 82
insects, 66–7

adaptation to environment, 91
navigation by, 68–9
Institute of Contemporary Arts,
2–3
insulin, 30
intelligence, 132
artificial, 12–13, 103–14,
128–30, 132–3
improvement of, 127–30
race and, 35
see also artefactual intelligence
invisible women, 181–4
IQ/race debate, 35
see also intelligence

Japanese Fifth Generation
Project, 13
Jordanova, L., 174
Joseph, Sir K., 184, 194

Kamin, L., 12
Keller, E.F., 174, 176, 187
Kerridge, S., 52
Kropotkin, P., 12, 87

labour
division of *see* production and
reproduction
double burden of women,
191–2
market and education, 190
Lamarck, J.B., 40
legal constraints *see* regulations
Lewontin, R., 12, 78
lichens, 84–5
life, duration of, 151
Lifton, R.J., 97
limitless vista of biology, 8, 19–25
limits of science
Rose, S. on, 8, 26–36
Watson on, 19–25
Taylor on, 161, 166–7
liquids, flow of, 56–8
logic versus history, 47–51
Lorenz, K., 65, 89
Lumsden, C.J., 92

McClintock, B., 187
McCulloch (scientist), 126
McKeown, T., 141, 142, 143, 150
machines
and brains, analogy between,
115–18
increase in use and costs of,
157–60
minds and meaning and, 13,
131–3
see also computers; technology
macromutation, 45
magnetism and electricity *see*
electricity
managerial responsibilities,
increasing, 157, 159
marginalization of women, 177
Marr, (scientist), 126
marriage, 95
Marxism, 187
see also socialism
material limits to science, 27–30
Maxwell, J.C., 161–2, 165
Maynard Smith, J., viii, 10, 87
on natural selection, 9, 11,
39–46
meaning, minds and machines
and, 13, 131–3
meaningful life, 24–5
mechanics, 164–5
Medawar, J.S., 74, 76
Medawar, Sir P., 1, 2, 3–4, 74, 76
medicine *see* health
Mendel, G.J., 47
Merchant, C., 186
midwifery, medicine's
appropriation of, 171–2
military research, 28–9, 184–6,
187
minds, machines and meaning
and, 13, 131–3
molecular biology, 29
see also genetics
Monod, J., 24
Morata, G., 52
morbidity rates, 146–7
morphology/form, 41–4, 48–51

fields and, 57–8
genes and, 51–3
laws of, 10
of matter, various, 164
novelty, emergence of, 59–60
process, 54–5
reproduction and, 54–6
see also homology
Morris, D., 65
mortality, 139, 140–1
explanation of, 141–6
perinatal, 34, 141, 143
mutation, 40, 44, 45, 51–2
see also transmutation

natural selection, 10–12
correct, 39–46
and emergence of novelty,
59–60
environment and, 9
levels of, 82–6
and self-interest, 66
structuralism versus, 9, 39–46
see also Darwin; evolutionary
theory
nature, forces united, 161–6
Navier-Stokes field equations,
56–7, 58
Necker Cube analogy, 81
Needham, (biologist), 29
neo-Darwinism *see* Darwinism
neuroscientists and brain, 125–7
neutrality
of science, claimed, 171, 176
of sociobiology, 62, 78
neutrino, 167–8
new science, towards, 14, 161–8
problems in, 14, 157–60
see also health; women
Newton, Sir I., 27, 48, 57, 165
Nixon, R., 8, 30
non-infectious causes of death,
144–6, 148, 151–3, 155–6
novelty, emergence of, 59–60
nuclear force, 162
nursery provision, 192–3

Oakley, A., 172
objectivity of science, 176
obstacles to development of
feminist science, 175–7
obstetrics, 155
Odell, F., 55
old age *see* elderly
Oosawa, F., 52
opportunities for women in
science, 195–6
optimism, 2, 8
organisms and fields, 56–9
organizational identity analogy,
117–18
Oster, G.F., 55
Oyama, S., 90

Pandemonium (computer
program), 107–9, 111–13, 128
Parsons, J.E., 175
Partridge, L., 90
patriarchy, capitalist, science
under, 183, 184–7
Pauli, W., 167–8
pentadactyl limb, 10, 48
Percival, Dr, 147–8
perinatal mortality, 34, 141, 143
Peto, R., 145
phenotypes, 10, 11, 84
philosophy of science, 19
physics
accelerators, particle, 28, 162
machines for, 158–9
spiral flow, 56–8
transmutation of elements, 49
unification of forces, 161–6
Piaget, J., 176
Picasso, P., 2
Pitts, (scientist), 126
Planck, M., 158
Platt, Lady B., 181
poetic analogy, 116
political response of doctors, 149
politics, 6, 87
see also sociobiology
pollution, 152
Popper, Sir K., 1, 4

prescription, social, and
biological evolution, 92–6
prevention of disease, 138, 150,
154
production
and reproduction, differences
and relationship between,
172–4, 175, 177–8, 189–90
research for, 185–6
profit *see* production
public
health, 148, 149–50
image of science and
technology, 3
punctuated equilibrium, 10

quantum mechanics, 164–5

race
and IQ debate, 35
science education and, 183
radical science movement, 8
radioactivity, 162
rational morphology, 49
Ravetz, J.R., 158–9
Reagan, R., 24, 160
reciprocation, 85
recombinant DNA, 20, 21–2, 25
reductionism, 7–8, 11
Dawkins on, 72–6
Rose, S. on, 29, 31–3
Taylor on, 166
regeneration, 53
regulations/laws
incest, 70–2
science and, 20–1
religion, 24
reproduction, 53–6
see also production
Rescher, N., 157, 158, 159
right, political, 6, 87
Roberts, J.M., 52
Rockefeller, J.D., 29
Roosevelt, F.D., 27
Rose, H., viii, 8, 172
on women, 15–16, 179–96
Rose, S., 80

Dawkins on, 71–4, 77–8
on limits of science, 1–16,
26–36
Rousseau, J.-J., 159
Rutherford, E., 116

Safran, J., viii, 133
on artefactual intelligence, 13,
115–30
Sanger, F., 30
Saraga, E., 170
Sayers, J., viii, 170
on feminism, 15, 169–78
schools *see* education
Science Policy Foundation, 2
Science of Science Foundation, 2
science *see* biology; evolutionary
theory; health; intelligence;
limits; machines; new
science; sociobiology;
women
Searle, J., 13, 132
seeing *see* computers, vision
selection *see* natural selection
selective attention, 122–4
self-fulfilling expectation, 88
self-interest and natural
selection, 66
selfishness, 5, 80–2, 83, 87
versus altruism, 66
'serial homology', 50
Sewall Wright, (biologist), 9
sex differences, 93
in mortality, 144–5
see also women
Sherrington, Sir C.S., 118
sickness, non-fatal, trends in,
146–7
Smith, Adam, 85
Smith, Alwyn, viii
on health, 14, 137–56
Snow, C.P., 1, 2, 3–4
social
behaviour, 64–8
conflict, 87–8
conformity, 94
co-operation, 80, 86–8

management of childbirth, 171–2
movements, new, and women, 187–9
organization and genetics, 5
organization of science *see* new science
prescription and biological evolution, 92–6
theory, critical, 186–7
welfare, research for, 185
socialism, 2
feminism and, 172, 173
opportunities for women and, 190
society, destabilization, 23–4
sociobiology
debate on, 11, 61–78
human politics and, 12, 79–99
arms race, 96–8
competition and co-operation and, 86–8
evolution, biological and social prescription, 92–6
genetic determinism, 88–92
selection, levels of, 82–6
selfish gene, 80–2
space–time
curved, gravity as, 163
dimensions of, 163–5
specialization in education system, 184, 194
spirals, 56, 57–8
Spurway, H., 41
stability of living matter, 42–3
stasis, evolutionary, 10
Stent, G., 15, 20
strangers, attitudes to, 97
structural analogy, 117–18
structuralism versus selection, 9, 39–46
surgery, 154
survival, 81, 83, 84, 87
value, 69–70
symbiotic co-operation, 84–5

Taylor, J., viii

on new science, 14, 161–8
technology *see* computers; machines; science
Teilhard de Chardin, P., 73
Tetrahymena, 52, 54, 57
tetrapod limbs, 59
Thatcher, M., 160
theories, discarded and replaced, 165
Thompson, Sir D'Arcy, W., 49
Thorpe, W.H., 89, 91
time
demands on, 191–3
space and, 163–5
Tinbergen, N., 65, 67
Trainor, L.E.H., 55
transmission, genetic, 84
transmutation of elements, 49
Truman, H., 27
Turing, (scientist), 126

unemployment, 153, 193
unification of forces of nature, 161–6
universe, beginning of, 166
universities *see* education

vaccination and immunisation, 142, 143
van den Berghe, P.L., 93
variation, 82
patterns of, 41–2
velocity field, 56
Verrell, P.A., 52
vertebrates, 42, 43–5, 50
Virchow, R., 150
vision *see* computers
Volta, Count A., 117
Von Guericke, (scientist), 117

'W and Z bosons', 162
Waddington, C.H., 29, 92
Wall, P., viii, 131–3
on artefactual intelligence, 13, 115–30
Wallace, A.R., 23
Walsh, (scientist), 117

war, 96–8
Watson, J., viii, 26, 27, 30, 31, 76, 170, 179, 187
on biology, 8, 14–15, 19–25
Weaver, W., 29
Weismann, A., 47
Westermarck, E., 94
Wilkins, M., 8
Williams, N.E., 52
Wilson, E.O., 11, 12, 60, 77, 88–9, 91–2, 170
Wilson, H., 2, 27
WISE *see* Women into Science etc.
women, 15–16, 169–96
absent and invisible, 181–4
capitalist patriarchy, science under, 184–7
education, science, 181–4, 194–6
reconceptualizing, 193–5
feminism, 6, 169–78, 187–8, 190
division of labour and new vision of science, 189–90
and socialism, 172, 173
opportunities, 195–6
social movements, new, 187–9
time, demands on, 191–3
Women into Science and Engineering, 181, 190
Woodger, J.H., 29
Woolf, V., 179, 180
World Health Organization, 136
Wrangham, R.W., 87